THE MAKING OF A CULTURAL LANDSCAPE

Heritage, Culture and Identity

Series Editor: Brian Graham,
School of Environmental Sciences, University of Ulster, UK

The Making of a Cultural Landscape

The English Lake District as Tourist Destination, 1750–2010

Edited by

JOHN K. WALTON

IKERBASQUE, Instituto Valentín de Foronda, University of the Basque Country UPV/ EHU, Vitoria-Gasteiz, Spain

JASON WOOD

Heritage Consultancy Services, Lancaster, UK

Routledge
Taylor & Francis Group

LONDON AND NEW YORK

First published 2013 by Ashgate Publishing

2 Park Square, Milton Park, Abingdon, Oxon OX14 4RN
711 Third Avenue, New York, NY 10017, USA

Routledge is an imprint of the Taylor & Francis Group, an informa business

First issued in paperback 2016

British Library Cataloguing in Publication Data
A catalogue record for this book is available from the British Library

The Library of Congress has cataloged the printed edition as follows:
Walton, John K.
    The making of a cultural landscape : the English Lake District as tourist destination, 1750-2010 / by John K. Walton and Jason Wood.
        pages cm
    Includes bibliographical references and index.
    ISBN 978-1-4094-2368-3 (hardback) 1. Tourism--England--Lake District--History. 2. Cultural landscapes--England--Lake District--History. 3. Lake District (England)--Social life and customs. I. Title.
    G155.G7W38 2013
    910.942--dc23

                                                                                    2013000861

ISBN 978-1-4094-2368-3 (hbk)
ISBN 978-1-138-24625-6 (pbk)

Contents

List of Figures and Tables

List of Figures

List of Tables

List of Contributors

David Cooper Senior Lecturer in English Literature, Manchester Metropolitan University, UK

Susan Denyer World Heritage Adviser, ICOMOS (International Council on Monuments and Sites), Paris, and Secretary, ICOMOS-UK

Keith Gregson Freelance writer, musician and historian, Sunderland, UK

Melanie Hall Associate Professor and Director of Museum Studies, Boston University, USA

Keith Hanley Professor of English Literature, Lancaster University, UK

Mike Huggins Emeritus Professor of Cultural History, University of Cumbria, UK

Adam Menuge Senior Investigator, English Heritage and Course Director, MSt in Building History, University of Cambridge, UK

Sarah Rutherford Freelance historic environment consultant specialising in conservation plans for designed landscapes, Buckinghamshire, UK

John K. Walton Ikerbasque Research Professor, Instituto Valentín de Foronda, University of the Basque Country, Vitoria-Gasteiz, Spain

Jonathan Westaway Research Impact Manager and Research Fellow, University of Central Lancashire, UK

Angus J.L. Winchester Professor of Local and Landscape History and Director of Victoria County History Cumbria Project, Lancaster University, UK

Jason Wood Director, Heritage Consultancy Services, Lancaster, UK

Foreword

Being given this book to read was a bit like finding myself a child in a sweetshop: for any advocate of landscape history and devotee of the Lake District, it is a must read.

The bulk of the book is the elegantly told story of the central role the Lake District has played in the evolution of the British preoccupation with and love of landscape. There is a strong narrative history of the evolution of the landscape and the pivotal roles played by luminaries such as Wordsworth, Coleridge and the National Trust's founders in articulating the special qualities of the Lake District. A crisis point was reached as it faced the pressures of the twentieth century: a largely pastoral landscape threatened by industrialisation and tourism which has come, eventually, also to hold the key to its future.

On the way we are treated to intriguing sub-stories, ranging from the emergence of the villa, the role of sport to the evolution of tourism and the riveting story of mountaineering and fell-walking. Furness Abbey and Claife Station are illuminating examples of the changing faces of tourism. The book ends with a reflective essay on Millom, part of Cumbria's industrial history neglected by the aesthetes.

I found myself drawn time and time again to the central preoccupation of the book: how can a complex cultural landscape like the Lake District be protected yet still evolve? Can its possible designation as a World Heritage Site with its concept of Outstanding Universal Value reconcile the competing tensions of people, beauty, nature and the economy?

So far, it is clear that in spite of many attempts we have not succeeded in doing so. This book will surely help us better understand the multi-dimensional nature of the Lakes and chart a better future.

Dame Fiona Reynolds DBE
Master of Emmanuel College, Cambridge and
former Director-General of the National Trust

Preface and Acknowledgements

This collection of essays brings together leading writers from a wide range of disciplines to explore the tourism history and heritage of the Lake District and its construction as a cultural landscape from the mid-eighteenth century to the present day. The book provides a commentary on the relationships between history, heritage, landscape, culture and policy that underlie the activities of the National Park, Cumbria Tourism and the proposals to secure the recognition of the Lake District as a UNESCO World Heritage Site. It explores key themes in the Lake District's history and identity, brings the story up to date and looks at current issues in conservation, policy and tourism marketing. The book is timely, in view of the World Heritage Site aspirations, and is intended to help scholars to exploit new sources and develop new insights into the history of tourism in Britain and beyond.

With or without World Heritage Site designation, and for whatever mixture of motives it might be pursued, the Lake District will continue to appeal to a complex mixture of cultural values and preferences, and to generate controversy over the appreciation, management and exploitation of its tangible and intangible resources, and the relationships between them. These issues are developed further, from a variety of standpoints and assumptions, in the chapters that follow. No 'party line' has been imposed on the contributors, and the expression of contrasting and contradictory views is to be expected. The essays constitute neither a seamless web nor a unified, coherent argument, but an accumulation of evidence and perspectives from which readers will draw their own informed conclusions.

We would like to take this opportunity the thank all of the contributors for their enthusiasm and commitment. We also extend our gratitude to various individuals and organisations for kindly granting permission to reproduce certain images and quotations: Susan Steinberg; Rowland Hart Jackson; Stephen Fullard; Pamela Woof; the Wordsworth Trust; Longfellow House – Washington's Headquarters National Historic Site; the National Trust; Country Life; English Heritage; the National Archives; Cumbria Archive Centre, Kendal; Cumbria Archive and Local Studies, Barrow; Millom Discovery Centre; Cumbria Tourism; and David Higham Associates. That the book appeared at all is due to the patience and dedication of all at Ashgate Publishing.

John K. Walton
Jason Wood
Lancaster

PART I
Lake District History and Identity

Chapter 1

The Lake District Landscape: Cultural or Natural?

Susan Denyer

If farmers left the Lake District valleys, allowing scrub to return to the high fells and fields to revert to un-drained bogs, would tourists still flock to the area?

The Lake District landscape has inspired and attracted writers, artists, scientists, political commentators and tourists for 250 years, and continues to be seen as a spiritual heartland for many visitors from the UK and further afield. Within the physical landscape, what is it that delivers the awe, the visual harmony, the profound sense of place and the notions of beauty that excited the early tourists, inspired writers, artists and conservationists, and is still a factor among many visitors today? Is the essence of the Lake District something static related to its accidental geography of mountains and lakes, its natural-ness, or is it related to the way the natural landscape has been shaped over time by a complex association between people and their environment, and thus to its cultural qualities? And what is its value? It is no exaggeration to say the Lake District has had a profound influence on the way landscape in general is perceived and conceptualised. Threats to its integrity prompted conservation battles in the nineteenth century to 'save' the Lake District that had in turn a significant influence on the creation of the conservation movement in the UK. Is the Lake District landscape, as Wordsworth memorably wrote, a 'sort of national property'?[1] Indeed, does it have a potentially global value in terms of beauty, influence and what it stands for? And if global, how might world-wide recognition sit alongside the way farmers and other members of local communities bestow their 'local' value? Whose responsibility is it to ensure its value is sustained for future generations to enjoy? How can the landscape be maintained in such a way as to conserve the very things that are the source of its beauty and harmony, and attract so much praise and attention? Can tourists be responsible agents in the supportive role Wordsworth envisaged? He foresaw them testifying to the value of, and the need to protect, the simplicity and beauty of a landscape created by generations of farmers. Even more practically, Beatrix Potter saw them contributing to support farmers' incomes:[2] is this still realistic, given the radically changed political climate since her death in 1943?

---

1   W. Wordsworth, *Guide to the Lakes*, preface by S. Gill (London: Frances Lincoln, 2004), p. 93.

2   S. Denyer, *Beatrix Potter at Home in the Lake District* (London: Frances Lincoln, 2000), p. 130.

From 2001 to 2012, a collaborative group of people preparing a World Heritage nomination bid for the Lake District found these questions challenging to answer. Their work has highlighted the complexity of the on-going interaction between people and the environment over time that such a landscape reflects. The Lake District is not just a collection of farming enterprises; it is a landscape that has been enhanced by associations with writers, artists and tourists. The Arcadian agro-pastoral landscapes that visitors 'discovered' in the eighteenth century have now been overlain in places with trees, gardens and villas, a process originally intended to enhance their Picturesque qualities. The landscape also reflects the success of nineteenth-century ground-breaking conservation battles and more recent tussles between those who vie for control of its resources, those who visit and those who use the landscape for agriculture. Perhaps most fundamentally of all it now also reflects a tussle between culture and nature in terms of how it is perceived and managed. Work on the potential Lake District bid has provided the opportunity for both local and wider communities to consider these contested concepts. It has prompted a wide-ranging debate on what value the Lake District landscape might be said to have at a global level, through comparing it with other similar landscapes around the world; and on what type of governance might be appropriate to sustain its value. Fundamental to this work has been the notion of 'cultural landscapes'. In World Heritage terms, UNESCO defines cultural landscapes as the 'combined works of nature and of man' that are 'illustrative of the evolution of human society and settlement over time, under the influence of the physical constraints and/or opportunities presented by their natural environment and of successive social, economic and cultural forces, both external and internal'.[3] Interestingly, the development of this definition, adopted in 1992, was actually prompted by two previous World Heritage bids put forward for the Lake District in the 1980s. The first submission was for a 'mixed natural and cultural property' while the second was for just a 'cultural property'. Neither bid was successful. The new category of 'cultural landscape' opens the door to the possibility of the Lake District being re-nominated for World Heritage Site status. It is now clear that only nomination as a cultural landscape stands any chance of success.

Apart from places of pristine nature, in one sense much of the world might be seen as cultural landscapes, since they reflect varying degrees of human interaction. World Heritage cultural landscapes are those areas that in some way manifest exceptional interactions, or exceptional outcomes of those interactions over time. Only landscapes that stand out over and above superficially similar places can be said to demonstrate Outstanding Universal Value (OUV), and thus be eligible for inscription as World Heritage Sites. The notion of OUV is at the heart of the World Heritage Convention.[4] Identifying such exceptional cultural landscapes requires

---

3 *Operational Guidelines for the Implementation of the World Heritage Convention* (Paris: UNESCO World Heritage Centre, November 2011), para 47.

4 *Convention Concerning the Protection of the World Cultural and Natural Heritage (known as the World Heritage Convention), 1972*, Article 11.2; see http://whc.unesco.

definition of their constituent elements: the environment that provided the raw materials, the processes that have shaped them over time and the precise outcomes of these interactions. What is also needed is a clear understanding as to how these might be read and understood within the landscape.

In terms of the 'raw materials', its geomorphology, the Lake District can be seen as a discrete area of small mountains rising out of the sands of Morecambe Bay to the south-west, the Irish Sea coast to the west, the Howgills to the east and the fertile plains elsewhere. It is divided into narrow valleys, many with lakes framed by fells that, as Wordsworth said, diverge like the spokes of a wheel from a hub somewhere between Great Gable and Scafell.[5] These mountains and lakes combine to provide an extraordinary diversity of valley forms, reflecting the complex geology of blue and green slate, granite and sandstone with slanting intrusions of grey limestone. The Lake District is at once diverse and coherent, with the whole encompassing little more than 800 square miles (or just over 2000 square kilometres).

How has this framework of mountains and valleys been shaped by people over time? The main interactions that have contributed to how the Lake District looks today, in terms of the disposition of its settlements and the patterns of its fields, can be broadly groups into four processes: agro-pastoralism, picturesque improvements, conservation battles and lastly, over the past 70 years or so, grants, subsidies and government and European policies. And these are neither sequential nor independent of each other as will be shown.

First, since at least the twelfth century and still persisting, the main process has been agro-pastoralism, communities gaining their livelihoods from a combination of growing arable crops in valley-bottom fields and the grazing of animals, mainly sheep and cattle, on the open fells. Of course these activities were supplemented by woodland activities, charcoal burning and the grazing of pigs and by industrial activities such as mining and smelting ores, but it was the socio-economic-cultural

---

org/en/conventiontext. The World Heritage Convention is significant for the way it links together the concepts of nature conservation and the preservation of cultural properties. It recognises the way people interact with nature, and the fundamental need to preserve a proper balance. Inscription on a World Heritage List requires acceptance by the UNESCO World Heritage Committee (elected by representatives of countries that have ratified the Convention) that the site deserves protection for the benefit of all humanity and transmission to future generations and unequivocally demonstrates 'Outstanding Universal Value'. In the Convention's early days, properties had to be nominated either for their cultural or for their natural values. But by 1984, the Committee had noted that in many countries there were landscapes that were not truly natural as they had been modified by people and that sometimes this interaction had created ecologically balanced, aesthetically beautiful and culturally interesting landscapes. A UNESCO Task Force considered in 1987 how these could be recognised under the Convention, but it was not until 1992 that a solution was found and the Committee adopted criteria for what came to be known as 'cultural landscapes': those that were neither purely natural nor purely cultural but rather a fusion between the two.

5   Wordsworth, *Guide to the Lakes*, p. 42.

parameters of agro-pastoralism that are reflected in the layout of the valleys as we see them today. Agro-pastoralism is a livelihood that has had a powerful influence on shaping many of Europe's landscapes and those of the wider Mediterranean area (and indeed large parts of the rest of the world). Many of the great monasteries supported this system, some reviving water management arrangements that had been first developed by the Romans. Today it is a way of life that is disappearing fast and with it the dramatic landscapes that it created. The Lake District is one of the last remaining examples in the UK and there are few comparators in Europe, one being The Causses and The Cévennes, Mediterranean agro-pastoral Cultural Landscape, France, inscribed on the World Heritage list in 2010.[6]

Just how agro-pastoralism shaped the landscape of the Lake District has been revealed in surveys of both physical and documentary evidence carried out by the National Trust and other organisations since 1981.[7] These have mapped many of the heads of the main Lake District valleys in a way that allows an understanding of their precise development by agro-pastoral communities and of what survives on the ground to reflect that development. These very detailed physical surveys mapped every field boundary, every building and all the extant archaeological evidence to present a powerful picture of the way the valleys have been shaped by people, particularly since the twelfth century. For the most part, their history over the past eight centuries can be read almost as a narrative from the surviving physical evidence. Many features were added to the landscape without necessarily erasing what went before. The basic framework of each settlement was created through communal activity. In such rocky terrain, no individual or nuclear family could have accomplished the massive tasks of moving rocks, digging channels, levelling fields and collecting, shifting and building the many thousands of tons of stone involved. Even when the basic infrastructure was in place, growing arable crops, grazing sheep and cattle, managing woodland and water courses and maintaining walls and hedges often involved more than one 'farm'. In the twelfth and thirteenth centuries, the valley-bottom fields were divided into strips, allocated to the surrounding farms. These fields were separated from the common, grazed fells above by a large all-enclosing 'ring-garth' wall. Later subdivisions amalgamated strips to create smaller fields. From the seventeenth century onwards, the lower slopes of the fells, often the best areas for grazing cattle, were gradually enclosed as 'intakes'. Their stone walls stretched up the fells, leaving 'outgangs' (walled paths) between to allow stock to pass up to the fell grazing. And as these changes were occurring over three or four centuries,

---

6   Details of The Causses and the Cévennes, Mediterranean agro-pastoral Cultural Landscape inscription can be found on the UNESCO World Heritage Centre website: http://whc.unesco.org/en/list/1153.

7   S. Denyer, 'Buildings in the Landscape', in D.M. Evans, P. Salway and D. Thackray (eds), *'The Remains of Distant Times': Archaeology and the National Trust* (Woodbridge: The Boydell Press, 1996), pp. 153–61; S. Denyer, *Lake District Landscapes* (London: The National Trust, 1994).

the woodlands that had clothed some of the upper fells gradually retreated, as settlements increased. Field divisions were mostly stone walls, some stout such as at Wasdale Head (where the huge quantity of boulders washed down by the becks is assembled into walls). Elsewhere, the field boundaries might be hedges, as in part of the once-communal fields between Buttermere and Crummockwater. With the exception of Conistonwater (where common fields and the ring-garth have not been delineated precisely), every valley surveyed reveals a similar evolutionary pattern of communal activity over time. The nucleated farmsteads were normally sited on the boundary between the enclosed arable fields and the grazed fell-land rising above or between the common fields. Sometimes farmsteads were arranged like a necklace around the edge of the arable land, as in Great Langdale and Borrowdale. At other times they were clustered together between two common fields, as in Watendlath and Buttermere. In some valleys they formed a 'street', such as the Row at Wasdale Head.[8] Following several centuries when the number of farmsteads increased, after the sixteenth century there was a decrease as holdings were amalgamated (Figures 1.1 – 1.3).

Crucially, the surveys reveal close and dependent relationships between the number, size, distribution and productivity of settlements and the management of water resources. Settlement of much of the low-lying flat land in valley-heads was only possible if the water that flowed down the fells and across the valley floors was carefully managed in terms of route and flow. The natural form of the valley-heads was mainly bog, with water meandering across it. The creation of communal fields and the building of farmsteads could only happen if communities constructed drains and managed these and the becks (streams) communally, clearing them of boulders on a regular basis. Until comparatively recently, such activity was still being undertaken at Wasdale Head, when all the community participated in 'boon' days to keep the channels open and prevent inundation of fields and buildings.[9] Revealingly, such approaches have counterparts in many mountainous areas of the world where the taming of water and regular maintenance of water courses is an absolute pre-requisite for sustainable communities. Just enough needs to be dredged to keep the water flowing, but going too deep negates the operation and purpose of the structures and risks destroying them altogether. For example, in the third century BC the Dujiangyan Weirs were constructed across the Min River in China as a basis for an irrigation system that allowed agriculture to flourish across the vast Chengdu plains of western China. The structures and systems are still in use today, over 2,200 years later. Within the water channels stone horse-shaped markers were buried. Every year, in the dry season, farmers dredged the water channels. The dredging would be over as soon as the horses were reached.[10] In the Lake District

---

8   Denyer, *Lake District Landscapes*, pp. 26–7.
9   Personal communication.
10   'Taming the Floodwaters: The High Price of Massive Hydraulic Projects', *China Heritage Newsletter*, 1 March 2005 (China Heritage Project & the Australia National University).

Figure 1.1   The development of the Watendlath valley landscape, east of Derwentwater, *c.*1350

Note: By this date the main elements of the landscape are in place. The farm houses form a nucleated group near the tarn. Nearly all the remaining flat land in the valley bottom has been enclosed by stone walls to form three fields – one around the tarn, a second behind the houses and a third alongside the beck to the north. These fields were divided into strips and managed in common for arable cultivation and as hay meadows. Above the fields the steep valley sides were clothed in dense woodland, and above the woods the fell land was open and managed as communal grazing for sheep, cattle and horses.

Source: © National Trust.

there were no stone horses but instead a wealth of communal knowledge that guided the sustainable management of essential, communal resources such as water.

Likewise collective management of the common resources of the high fells was and is a crucial part of the process that shaped the landscape. Even after some partial enclosure of fells in the nineteenth century, the Lake District commons represent the largest concentration of common land in Britain, and possibly the largest in Western Europe too. This commonality was not just a legal concept; it was intimately bound up with land management, animal husbandry and human culture. 'Hefted sheep' – those that always return to the place where they were weaned – play a crucial role in this collective management regime. Flocks distinguished by their smit marks were tied to specific farms but mixed freely on the fells. Most sheep breeds do not exhibit this property but it is a remarkable characteristic of the Herdwick breed whose distinctive profile, white face and

**Figure 1.2	The development of the Watendlath valley landscape, east of Derwentwater, *c*.1600**

Note: Around 250 years later, the overall pattern remains unchanged but some of the common fields have been subdivided into small fields attached to specific farms. Part of the lower slopes of the woodland have been cleared and enclosed by stone walls as 'intakes' for cattle grazing, while some of the higher woods have been thinned to become wood pasture. The number of houses has increased and the fells still remain communally managed.

Source: © National Trust.

grey fleece still forms such an evocative sight on the highest Lake District fells.[11] Sheep and common land are inextricably linked by a complex socio-economic system that has persisted for centuries. In turn it relates to the relative freedom of the tenant farmers in their comparatively remote valleys, and a system of self-regulation formerly through Manorial Courts and more recently through commoners' organisations. By the seventeenth and early eighteenth century, this distinctive farming system had reached the peak of its prosperity. Many houses had been rebuilt with a second storey, embellished with oak panelling and fitted with grand carved oak cupboards, and the yeoman farmers – which Wordsworth said were 'small independent proprietors of land here called statesmen' – were a proud and seemingly almost independent community.[12]

11	G. Brown, *Herdwick Sheep and the English Lake District* (Kirby Stephen: Hayloft, 2009).

12	*The Letters of William and Dorothy Wordsworth: Volume I. The Early Years 1787– 1805* (Oxford: Clarendon, 2000), p. 314.

**Figure 1.3   The development of the Watendlath valley landscape, east of
Derwentwater *c.*1850**

Note: A further 250 years later and the overall pattern still persists but by now all of the
common fields have been divided by stone walls into smaller fields; many more intakes
have been created and most of the woods thinned to become wood pasture. The number of
houses has reduced to a third of the number in around 1600 AD and most have been rebuilt
or enlarged. The fells have been enclosed and allocated to individual farms.
Source: © National Trust.

Such was the physical and social landscape that early visitors to the Lake
District encountered in the second half of the eighteenth century. Wordsworth in
his *Guide to the Lakes* looked back nostalgically to a period roughly around the
time of the early tourists, which he saw as a golden age of the 'perfect Republic of
Shepherds and Agriculturalists', a sort of egalitarian society without noblemen or
squires that supported each other and where farmers 'had a consciousness that the
land, which they walked over and tilled, had for more than five hundred years been
possessed by men of their name and blood'.[13]

By the end of the eighteenth century, this Lake District landscape was about
to be shaped by new processes. Its beauty became a focus for wealthy tourists
who, denied access to the Grand Tour in Europe through unrest in France, began
to turn their attention to mountain areas of England, Wales and Scotland where
they found an alternative, particularly to the Alps. The new tourists brought visual
notions of 'Arcadia', as depicted by painters who alluded to Classical traditions

---

13   Wordsworth, *Guide to the Lakes*, pp. 74–5.

**Figure 1.4   An engraving of Thomas Smith's view of Derwentwater from
Crow Park, published in 1767**
Source: © Wordsworth Trust.

of Greece and Rome. Well-watered valleys, oak woods and the herding of sheep
and cattle all fitted rather well into Arcadian concepts, which were transferred in
their minds to the Lake District.

Artists quickly followed the tourists. They painted what the tourists wanted
to see, in an idiom that could be understood by that small elite audience schooled
in classical landscape traditions of seventeenth-century Rome. For instance,
Derwentwater quickly took its place as a passable alternative to the Alban hills
near Rome, helped by artists such as Thomas Smith who obligingly depicted it
as an almost circular lake with woods cascading into the water from surrounding
hills, and in the foreground cattle peacefully grazing (Figure 1.4).

By the 1780s and 1790s there was a stream of artists coming to the Lake
District to paint this northern Arcadia, as a way of establishing their reputations.
The artists were joined by poets such as Wordsworth, Blake, Coleridge, Byron,
Shelley and Keats. Later known as the Romantic poets, they, and particularly
Wordsworth, created poetry to express ideas related to perceptions of landscape:
intuition over reason; pastoral over urban; and the reconciliation of man
and nature.

As the century progressed, the trickle of visitors increased. In 1778 Father Thomas West produced his *Guide to the Lakes*.[14] In it he defined 'Stations', rather in the manner of 'stations of the cross' at pilgrimage sites. His secular stations were sited around a lake where the views could be seen as forming a perfectly proportioned picture. Station number one on Windermere was on the eastern slopes of Claife Heights. There a small building was constructed through whose variously coloured-glass bay window could be seen framed views that allowed understanding of how the lake scene might look in the four seasons, at midnight and in a thunderstorm.

The idea that certain landscapes embodied perfection in terms of the relationship between their various elements quickly led on to the idea that landscapes that did not have all these elements might be 'improved' in order to acquire them. Thus began a new phase in the shaping of the landscape when 'off-comers', with ideas of beauty and harmony, changed from being observers to 'interveners' who adopted the landscape, buying land which they 'enhanced' with villas, gardens and woodland. The aim was to make picturesque improvements to their 'natural' surroundings. In one sense all of this could be seen as a complete re-moulding of the landscape. However, although some valleys such as Windermere underwent substantial change, many others were touched more lightly. The changes did not completely overwhelm the farming landscape and can be read as a picturesque overlay, in some valleys confined to a few, well-placed, non-native trees. This imprint of villas, designed landscapes, arboreta and woodlands broadly took place between the 1770s until the 1920s. More than any other lake, Windermere has been modified through the building of villas and the planting of trees and gardens. The first intervention was the construction of Belle Isle in 1774, a circular house on an island in Windermere. It was associated not only with ornamental planting on the island but with the afforestation of large areas of Claife Heights sloping down to the west shore of Windermere to provide a backdrop to the new house. From Brathay Hall at the northern end to Fell Foot at the south, a chain of distinguished villas arose along the east side of the lake.

In other valleys, the planting of trees had more impact than the building of villas. On Derwentwater the east shore of the lake was planted in contoured 'layers' for picturesque effect; the mainly oak woodland was framed with a row of beeches along the edge of the Lake and Scots pines along the skyline. On the west side, Lord William Gordon engaged in extensive tree-planting on his land, stretching from Fawe Park to Manesty. Along the west side of Ullswater over a period of about sixty years from the 1820s, the Howard family of Greystoke

---

14   T. West, *A Guide to the Lakes: dedicated to the lovers of landscape studies, and to all who have visited, or intend to visit the Lakes in Cumberland, Westmorland and Lancashire. By the author of the Antiquities of Furness* (London: Richardson & Urquhart and Kendal: W. Pennington, 1778). Father West suggests a number of 'stations' for some of the lakes: some are in numbered sequences; others are referred to in the text.

Castle planted many hundreds of thousands of trees in Gowbarrow Park[15] and in the late nineteenth century, a pinetum framing the nearby Aira Force waterfalls. Impressive arboreta were planted at Monk Coniston Hall, near Coniston, and High Close, Loughrigg. Further to the west, the valleys reveal fewer overlays but there are nevertheless some dramatic interventions such as around Wasdale Hall at the southern end of Wastwater. Facing across to the bare screes, and planted in the 1840s, the woods included many exotics grown from seeds brought back from the Eastern Mediterranean by Stansfield Rawson.[16] The re-introduction of Scots pines provided non-deciduous trees that were relatively easy to grow. Groups were planted for dramatic effect at the head of some lakes, as at Gatesgarth Farm, Buttermere, and around Blea Tarn in the Langdales. At Watendlath, their dark green foliage contrasts strikingly with the whitewashed facades of the farms. Wordsworth planted some at Rydal Mount, and many other houses, both grand and more modest, exhibit similar patterns.

By the early nineteenth century, such interventions began to elicit sharply divergent responses. Some people felt the 'enhancements' were to be welcomed, while others considered there were strictly finite limits to what was acceptable in modifying 'traditional' agro-pastoral landscapes. What was deemed acceptable intervention became the subject of intense debate. The interaction between people and their environment was becoming increasingly complex.

The third process that had a dramatic impact on the landscape was when people, both individually and collectively began to intervene to protect and conserve a landscape that they considered to be of value and was under threat. Certain interventions might be welcomed amongst the growing body of people who identified themselves with the Lake District. But some developed unease about anything that could be seen as detracting from the vision of Arcadia. This particularly applied to the felling of indigenous trees. When a stand of noble oaks was felled at Crow Park at the north end of Derwentwater there were protests, and a painter (commissioned by the Speddings of Whitehaven who bought the timber) captured what was seen as an emblematic moment.[17] And negative changes were not confined to trees. Wordsworth was a trenchant critic of some Picturesque interventions. He disliked the idea of selecting sites for new villas on the basis of their 'craving for prospects'[18] which necessarily meant they became highly visible in the landscape. He found planting without regard for the natural contours of the land abhorrent, all the more so when important views were blocked. Wordsworth did not condemn all such interventions. Rather, in his *Guide* he enunciated principles that would allow changes that respected the 'peaceful harmony of form

---

15 Catalogued in a notebook belonging to the Howard family, Greystoke Castle.

16 J. Martin, 'Wasdale Hall', *Trans Cumberland and Westmorland Antiq. and Archaeol. Soc.* new ser., 93 (1993), pp. 269–82.

17 Illustrated in G. Bott, *Keswick: The Story of a Lake District Town*, 2nd edn (Wigton: Rosley Books, 2005), p. 35.

18 Wordsworth, *Guide to the Lakes*, p. 80.

and colour'.[19] In a section entitled 'Changes, and Rules of Taste for Preventing their Bad Effects', he suggested interventions should be made in such a way as to become 'gently incorporated into the scenery of Nature'.[20] The book enunciated precepts for siting new buildings 'to harmonise with the surrounding landscape'; to respect local building traditions; and to reflect what he called picturesque garden design. He himself designed a house based on traditional materials and details (but not traditional scale), although it was never actually built. Later, not long before his death, he was prompted to consider that the castellated Wray Castle built on the edge of Windermere added a 'dignified feature to the landscape scenery in which it stands'.[21]

These processes of interaction that were shaping the Lake District landscape were poised to become yet more complex. The impact of increasing numbers of tourists started to threaten the very beauty that the early tourists had discovered, while at the same time the social structures of the farming communities that underpinned the landscape were beginning to weaken. Wordsworth's *Guide* appeared just as middle-class tourism to the Lake District was growing. The economic changes that fostered this tourism – the growth of towns, cities, manufacture and transport with the industrial revolution – had dealt a double blow to the Arcadian vision. As visitor numbers increased, population migration from the Lake District valleys to urban areas led to a decline in the number and success of small farms. There was also a loss of subsidiary industries, such as small-scale mining at this time, and an increase in mortgaging, with its risks. All this is particularly clear in the acute observations of Wordsworth, and his sister Dorothy, about the 'common' people and the way they worked the land and the stresses they were enduring. Those stresses were reflected in the farming landscapes too.

In *Michael, A Pastoral Poem*, written in 1800, Wordsworth writes about a shepherd, who had lived and worked all his life in Grasmere on the land he loved. 'Stout of heart, and strong of limb ... of an unusual strength: his mind was keen'.[22] His only son, Luke, is sent to be an apprentice in a town to help pay a family debt, but there falls into dissolute ways and goes abroad. Michael is left working alone and unable to finish building a sheepfold they had started together. After his death, the house and land are sold 'into a Stranger's hand', and eventually all that is left of the home is an oak tree and the unfinished sheepfold. These profound economic and social changes meant that when families were no longer able to maintain themselves on their small farms, their land was vulnerable to acquisition by 'off-comers'. Many small estates were newly created, with mansions erected out of the ruins of 'ancient cottages'. As Dorothy Wordsworth said, there would soon be 'only two ranks of

---

19   Ibid, p. 79.

20   Ibid, p. 83.

21   Sir B. Burke, *Visitation of the Seats and Arms of the Noblemen and Gentlemen of Great Britain*, vol. 2 (London: Colburn, 1853), p. 148.

22   D. Porter and M. Garner (eds), *William Wordsworth and Samuel Taylor Coleridge, Lyrical Ballads: 1798 and 1800* (Ontario: Broadview, 2008), p. 385.

people – the very rich and the very poor.[23] As well as writing about the plight of farmers, in poems such as *Michael* and the *Brothers* (on the Ewebanks of Ennerdale), Wordsworth also took to campaigning. In 1801, he wrote to Charles James Fox, then Leader of the Opposition in the House of Commons, asking for support for a 'class of men' that 'is rapidly disappearing'.[24] And, in 1824 he successfully opposed proposals to enclose the Rydal Commons.[25]

By the middle of the nineteenth century the trickle of tourists had turned into a flood, helped by the building of railways to Windermere and Keswick. Suddenly the Lake District was within reach of the rapidly growing populations of the North East, Lancashire and Yorkshire and beyond. Instead of just the occasional villa or garden, substantial hotels and settlements arose, as newcomers acquired land and invested in facilities for tourists. But there was also a shift in the way the landscape was perceived. Tourists were beginning to come for the peace and quiet of the countryside, fleeing from the noise and grime of mushrooming industrial towns. But they no longer wanted merely to look at perfectly proportioned views of the landscape: they wanted access to it. The open fells, some of which had recently been enclosed and were patterned with dry-stone walls, came to be seen as places that could give solace to town dwellers, and to have value for their ability to deliver exercise, enjoyment and recreation. The tourists wanted access to the lakes and fells at the same time as other off-comers blocked this access with grand lake-side houses or fences on fells. More railways were planned; the growing industrial towns needed water; and the Lake District was seen as a place to build reservoirs in order to supply this need. The interaction between farmers and outsiders, reflecting the impact of wider social and economic changes, was to lead to recurring conflicts. The value of agricultural land was beginning to become divorced from its agro-pastoral capacity.

Wordsworth campaigned for what he saw as the essence of the landscape, why it was of value and how its 'right development' should be approached. He was followed, by many other influential figures capable of mustering responses to address threats to the landscape they considered had a value for all. Their eloquence, leadership and successes had a profound impact on the way the landscape developed. If it were not for these campaigners, the Lake District would now be a very different place. The landscape as it is today can thus to a degree be read as a testimony to the idea of conserving places for the benefit of society as a whole. Although Wordsworth did not live long enough to see the conflict really develop between public and private interests, he saw enough to predict the possible demise of the rural way of life. It was left to others to come up with solutions. Six years after Wordsworth's death in 1850, the Keswick Footpath Preservation

---

23   P. Woof (ed.), *Dorothy Wordsworth: the Grasmere and Alfoxden Journals* (Oxford: Oxford University Press, 2002), p. 3, entry for 18 May 1800.

24   *The Letters of William and Dorothy Wordsworth* ..., p. 314.

25   H.D. Rawnsley, *Reminiscences of Wordsworth among the Peasantry of Westmorland* (London: Dillons, 1968), p. 27.

Association was founded, and it was soon doing battle with the new landowners against footpath closures. This was to become a recurrent theme.[26] In the late 1870s, there was a battle to stop Manchester Corporation building Thirlmere reservoir. Although unsuccessful, this battle changed the nature of disquiet against what were seen as undesirable changes to landscape. Firstly, national figures were drawn in to support a campaign that led to concerted protest and the formation of the Thirlmere Defence Association – the first national association for the protection of landscape. Secondly, and perhaps most important of all, the campaign was seen to legitimise the assertion that the aesthetic appearance of landscape and the recreation that it could deliver both had a value. As a newspaper editorial stated: 'The lake country belongs in a sense, and that the widest and best sense, not to a few owners of mountain pasture but to the people of England'.[27] The value of the Lake District landscape held by those who lived outside the area was acknowledged as being of relevance in determining its future development.

A key supporter of the campaign against Thirlmere was the artist, critic, social thinker and philanthropist John Ruskin (1819–1900) who in 1872 came to live at Brantwood, on Coniston Water. He took up the cause of conservation with much passion and vigour, and many of the national issues on which he campaigned are still valid today – town and country planning, green belts and smokeless zones. In the Lake District, through his support for campaigns against the closure of footpaths, the extension of railways, the building of reservoirs and creeping industrialisation, Ruskin began to consider the idea of positive intervention for landscapes of scenic beauty to safeguard them for the longer term. Building on Wordsworth's concept of the Lake District as a 'sort of national property' for all those who 'have an eye to perceive and a heart to enjoy', Ruskin suggested a form of national ownership for such areas.

While at Oxford, Ruskin had met Hardwicke Rawnsley (1851–1920). Rawnsley took up the post of the Vicar of Wray, near Hawkshead, in 1878 and then in 1883 he was appointed to the living of Crosthwaite Church near Keswick. Rawnsley adopted the ideas of Ruskin but gave them practical effect. No sooner had he arrived in Keswick than he founded the Lake District Defence Society. Under his encouragement, distinguished national figures joined such as Alfred, Lord Tennyson, Robert Browning and Ruskin himself, as well as many other members from the national Wordsworth Society. The threats facing the Lake District were not unique and to varying degrees were reflected in other areas of the countryside appreciated for their landscape value. By degrees an alliance developed nationally

26	W.J. Darby, *Landscape and Identity: Geographies of Nation and Class in England* (Oxford: Berg, 2000), p. 150–51.

27	J.D. Marshall and J.K. Walton, *The Lake Counties from 1830 to the Mid-Twentieth Century* (Manchester: Manchester University Press, 1981), pp. 207–14; H. Ritvo, *The Dawn of Green* (Chicago: University of Chicago Press, 2009); also

H. Ritvo, 'Fighting for Thirlmere – The Roots of Environmentalism', *Science* 300 (6 June 2003), p. 1511.

between society figures, academics and thinkers, and those working on the ground, such as Rawnsley, who shared the idea that only through some sort of benign ownership could these threats be addressed. This eventually led to the formation in 1895 of the National Trust created to buy land and own it for the benefit of the nation. The three founders of the Trust were Octavia Hill, a social reformer and Oxford friend of Ruskin's to whom he had introduced Rawnsley; Hardwicke Rawnsley himself; and Robert Hunter. In many ways Ruskin could be considered as an associate founder. Under the terms of the National Trust Act of 1907, land and buildings deemed to be of 'inalienable quality' could be purchased by the Trust under advantageous tax arrangements in return for an undertaking that such property would not be sold and would remain an asset 'for ever, for everyone'.[28]

The foundation of the National Trust provided the opportunity for it to purchase areas of the Lake District seen as being under threat. But in order to implement its remit, there was a need for there to be some sort of shared understanding of what the essence of the Lake District was and what parts of it needed protection. Although not precisely articulated at the time, there emerged an implicit agreement that the relationship between lakes, in-bye land, woodland, grazed fells and the built environment needed to be sustained in order to conserve the picturesque beauty of the landscape. An extension of the Windermere villa-building pattern to other shores of that lake, and to other lakes, would mean the loss of the Lake District's unique qualities forever. Such ideas are reflected in many of the Trust's earliest purchases, such as Brandelhow on the edge of Derwentwater, its first acquisition bought with funds raised through a public appeal in 1902, and other lake-side land threatened by development of villas and hotels. These acquisitions prevented the lakes being cut off from their surrounding fields, and thus protected the basic structure of the valleys. In seeking out places to protect, there was not a wholesale rejection by the Trust of landscape that had been 'improved' with trees and arboreta, as some interventions were acknowledged as having added to the landscape in a positive way. Aira Force and the nearby Gowbarrow Park, Ullswater (which had been up for sale for housing when purchased by the Trust) were two cases in point. Over time, the Trust acquired many other properties with picturesque dimensions, such as Monk Coniston Hall at the head of Conistonwater, Derwent Island House, Wray Castle overlooking Windermere, Blea Tarn between Great and Little Langdale, Allan Bank at the head of Grasmere, the head of Buttermere and the Wasdale Head Hall estate.

In the first 25 years of its existence, the National Trust's acquisitions in the Lake District show a keen appreciation of the harmony of interplay between farming and the picturesque, along with a determination to stop major development that would upset such relationships. It seemed that the 'pastoral and pure' views that Ruskin admired would be safe. But in the 1920s the threats began to shift from lake shores to the very heart of the Lake District valleys. As Wordsworth so

---

28 G. Murphy, *Founders of the National Trust* (London: Christopher Helm, 1987); M. Waterson, *The National Trust: the First Hundred Years* (London: National Trust, 1997).

graphically described in *Michael*, there had been a decline in the number of farms since the second half of the eighteenth century, as amalgamations took place. But during his lifetime this had been simply a change of scale, not a threat to farming itself. By contrast, the economic depression of the 1920s began to reduce the value of farmed land to such an extent that small plots were being bought for chalets, fields for housing development, and much larger tracts of land for planting commercial crops of trees, encouraged by the Forestry Commission established in 1919 to improve the supply of timber. The whole farming operation of the valleys was becoming vulnerable. Such a massive threat could only be addressed by a massive response. Fortunately some of the Lake District's supporters and adopted residents had the funds and the foresight to rise to this challenge. Such enlightened donors and benefactors allowed the National Trust to expand its land holding in the main valleys so farmers could continue farming in the traditional manner. Others purchased land themselves with the intention of making it over to the Trust on their death. Some did both. One of these was Beatrix Potter, the author and illustrator, who by this time was in the fortunate position of receiving substantial royalties from her 'Little Books'. She saved many farms in some of the best known valleys of the Lake District. Had she not intervened, the Coniston and Tilberthwaite valleys, Troutbeck and the Langdales would today not be farmed but would almost certainly consist of the remains of abandoned buildings surrounded by softwood plantations: the central Lake District would be unrecognisable.[29]

But ownership by the National Trust and other sympathetic landowners was still not enough. The threat of large-scale afforestation brought into sharp focus the need for more formal protection to sustain the Lake District landscape as an entity and give it a national status. By the early 1930s, the Forestry Commission had afforested huge swathes of Ennerdale. When it acquired land in upper Eskdale and Dunnerdale (the Duddon valley) in 1934, and a year later announced its plans to plant larch and spruce over around a third of it, opposition reached a crescendo. Cohorts of those who had adopted the Lake District as their home and many illustrious outsiders mounted a petition to reverse the plans and to put in place future protection from such large-scale afforestation. The main arguments in print focused on the openness of the fells and the freedom to roam that they provided, reflecting the huge interest in hiking in the 1930s, supported by such bodies as the Ramblers Association, the Youth Hostel Association and the Campaign for the Protection of Rural England. But one of the most memorable interventions on the importance of protecting the farming landscape came from the Revd H. H. Symonds who stated in his evidence to the Forestry Commissioners that the afforestation would wipe out the valley-head farms that were essential to sustain the Herdwick sheep and would 'see a proud people regimented out of existence by the invasive conifer'. The protests succeeded to an unprecedented extent. An agreement was reached that there should be no large-scale afforestation in the central Lake District. Admittedly, the 300 square miles to which protection

---

29   Denyer, *Beatrix Potter*, p. 128.

was granted was smaller than the 700 square miles that had been demanded. But the fundamental outcome was that wider national protection for the whole Lake District now had to be taken seriously.[30]

Impelled by promises of 'a better world for all' made by politicians to war-torn communities and beleaguered service personnel in the depths of the Second World War, the peace of the 1940s saw strong political support for action on the protection of the homeland, and attributes like mutual respect and natural beauty. One of the first initiatives of the 1945 government was to commission John Dower to advise on how the National Park ideal developed in North America in the 1860s might work in England and Wales. Dower had a wide vision of parks where landscape beauty was to be preserved, access provided, wildlife and buildings protected and established farming systems maintained. Although he also talked of 'beautiful and relatively wild country', his vision was very much in line with the idea of what much later came to be called cultural landscapes. Dower's 1946 recommendations led directly to Arthur Hobhouse's 1947 report which in turn prepared the legislation for the National Parks and Access to the Countryside Act of 1949, and the subsequent creation of twelve National Parks. The reality of National Parks was slightly different from Dower's vision. The Act brought in the idea of natural beauty that could provide opportunities for open-air recreation.

The Lake District became a National Park in 1951. Wordsworth's idea of a 'sort of national property' had become reality 101 years after his death. This momentous step defined the limits of the Lake District and captured some of the things for which it was to be valued. With the creation of the Lake District National Park Authority, it also raised for the first time the question as to precisely how this landscape was to be managed as an entity. And thus comes the fourth of the main processes that have shaped the landscape: management of the National Park to reflect national and later European agendas.

Up to the eve of creation of the Lake District National Park, it could be argued that the overall development of the Lake District reflected the views of farmers, wealthy off-comers, entrepreneurs, campaigners for access and residents who all believed passionately in the need to protect its landscape. The dynamic tensions between these groups, together with the moral and financial support of a larger constituency outside the area who shared their belief in the importance of the Lake District landscape, had shaped thinking over the previous 150 years. The creation of the National Park took the initiative from this loose consortium and opened up issues as to whether the agenda was to be set at local level or at national level, and to what degree different communities, such as farmers and tourists, would receive benefits.

---

30   H.H. Symonds, *Afforestation in the Lake District: a Reply to the Forestry Commission's White Paper on 26th August 1936* (London: J.M. Dent, 1936); I.O. Brodie, *Forestry in the Lake District: a Retrospective Look at H.H. Symonds' 1936 Book Afforestation in the Lake District* (Kendal: Friends of the Lake District, 2004).

In the intervening six decades, tussles about the way the landscape should be managed and who should benefit have intensified. Reflecting national policies and more recently international agendas too, there have been huge shifts in emphasis in management thinking and action. Whether protective designations, or the availability of grants and subsidies, most such external vectors have tended over time to result in more emphasis being placed on the natural rather than on the cultural values of UK landscapes in general and the Lake District in particular. In England, the National Park Act of 1949 encapsulated the tensions between culture and nature. It promoted access to beautiful areas without a specific recognition of the way those areas had been shaped by people, and it implied that upland areas – for all the early National Parks centred on the uplands – were somehow purely 'natural'. To a degree, this picture was modified by the Environment Act of 1995: this revised the statutory duties of National Parks so that conservation and enhancement of natural beauty, wildlife and cultural heritage and promotion of opportunities for understanding and enjoyment were now all explicit. But natural beauty was still seen as something separate from cultural heritage. Both these Acts enshrined the philosophy of habitat conservation and building conservation through site designation; they did not integrate nature conservation or building conservation with each other, or with the broader framework of rural policy. And perhaps most fundamentally Section 62 of the 1995 Act makes clear that if National Park purposes are in conflict, then nature 'conservation' must have priority. This was part of the Sandford Committee's recommendation in 1974, that enjoyment of National Parks 'shall be in a manner and by such means as will leave their natural beauty unimpaired for the enjoyment of this and future generations'. Accordingly, it is known as the 'Sandford Principle'.

The polarisation between nature and culture was made more acute in the 1970s and 1980s when farmers, in order to survive, were forced to seek a variety of grants and subsidies specifically for agriculture that took too little account of nature conservation. This led to increasing threats to Sites of Special Scientific Interest (SSSIs), such as bogs, rivers, lakes, butterfly habitats, etc. – whose designation had been allowed for in the National Parks Act – and to increased efforts to protect them from what was seen as hostile land-use. SSSIs tended to become protected islands within the farmed landscape. Further, the desirability of restoring woodlands to what were seen as their natural and semi-natural states also gained ground in order to make them more valuable as habitats. Exotic trees were felled including most of all those fine specimens planted in woods near Wasdale Head Hall in the 1840s.[31] The noble beech trees along the edge of Derwentwater were only just saved. Micro management of small 'natural' areas gained ground at the expense of sustaining the wider entity. And tensions developed between those who farmed the landscape and those who saw its natural aspects as being 'degraded'.

---

31   Martin, 'Wasdale Hall', pp. 269–82.

The introduction of the Environmentally Sensitive Areas (ESA) Scheme in 1987 brought European funds into the Lake District. The scheme offered financial incentives to encourage farmers to adopt agricultural practices that would safeguard and enhance landscapes of nature conservation or historic value in designated areas. The Lake District became an ESA in 1992, subsequently superseded by the Environmental Stewardship scheme with broadly similar objectives. The ESA scheme was valuable in bringing together cultural and natural interests within an economic framework. However, whereas the environmental benefits flowing from the scheme were on-going and dynamic (for example, improvements to species-rich grassland and hay meadows), the cultural benefits were linked to the repair of historic features such as walls and buildings and thus far more static. Nevertheless it proved to be a valuable catalyst that led to the conservation of over 500 farm buildings as well as many miles of stone walls. In turn, these activities have fostered traditional skills.[32] The ESA scheme also brought into focus the overall 'health' of the fells. Farmers who joined were required to reduce the number of grazing animals per hectare of fell in order to increase the biodiversity of the natural cover and reconstruct plant communities for public good. These obligations renewed tensions, as the restrictions were seen as severe, threatening the viability of some farms, leading to amalgamations and worse. In setting targets for stocking levels, the health of the cultural system that underpins the landscape was not seen as a consideration, rather only the health of plant communities.

In the wider world, 1987 was also the year the World Commission on Environment and Development defined and promulgated the concept of sustainable development. Its report, *Our Common Future* (also known as the Brundtland Report), recognised the environment not as something separate from development but closely linked to it by human emotions and human action. 'The environment is where we live; and development is what we all do in attempting to improve our lot within that abode. The two are inseparable', the report said.[33] But somehow in the Lake District and in other upland areas the gulf remained and indeed was widened by the Habitats Directive of 1992.[34] This European Commission instrument provides protection for designated animal and plant species and habitat types of European importance. It allowed certain areas to be protected at a European level, over and above national protection. Such areas further dislocated what should have been acknowledged as dynamic, functional links between the environment, people and development.

---

32   S. Denyer, 'Recording Farm Buildings in the Lake District', *Proceedings of a Conference on Recoding Farm Buildings* (London: RCHM, 1994).

33   G. Brundtland (ed.), *Our Common Future: The World Commission on Environment and Development* (Oxford: Oxford University Press, 1987), p. 45.

34   The Conservation (Natural Habitats, &c.) Regulations 1994; *The Habitats Directive: How it Will Apply in Great Britain* (London: Department of the Environment, 1995).

The Rio Declaration on Environment and Development of 1992 was launched to bring forward initiatives and goals on the basis of the Brundltand Report. Its Agenda 21, with ideas of sustainable and responsible stewardship of resources, could have been the catalyst for considering the way cultural landscapes reflect resilient and effective systems of management with the capacity to deliver wide benefits in terms of aesthetics, access and food. In the UK, though, Agenda 21 had a much narrower focus. The three pillars of sustainable development that it defined – social, economic and environmental – tended to have the effect of removing culture from considerations of landscape conservation and management. So-called 'sustainable environmental management' pushed the focus away from landscapes as coherent entities reflecting interaction between culture and nature, towards a comprehensive stewardship of natural resources. The subsequent Johannesburg Declaration on Sustainable Development in 2001 (and the 2002 Porto Alegre Agenda 21 for Culture) acknowledged the need to ensure culture was fully integrated into sustainable development as a fourth 'leg'. But this notion has had too little resonance in the UK. At an international level, however, the Rio Declaration has achieved notable success in prompting a reflection on how cultural landscapes might be considered for inscription on the World Heritage list.[35]

One episode that highlighted the lack of understanding of the Lake District as a cultural landscape was the outbreak of Foot and Mouth Disease (FMD) in 2001. Miraculously, FMD was hardly detected in Lake District farms but in the surrounding plains there were terrible consequences for farmers in terms of mass slaughter of sheep and cattle. The FMD outbreak showed starkly the apparent lack of central government awareness of the Lake District as a landscape shaped by cultural traditions over time. Letters from the National Trust and others urging the government to take measures to protect the Herdwick flocks in order to safeguard the landscape were not understood and were met with responses that failed utterly to appreciate the consequences of large-scale destruction of 'hefted' sheep. There was no sense of impending doom that this very distinctive upland landscape with its coherent, long-standing ways of managing the land was on the edge of complete catastrophe. One person who did understand the gravity of the situation was the veteran journalist Harry Griffin. He wrote just after FMD reached Black Hall Farm in the upper Duddon valley on 11 April 2001:

> If the scourge of foot and mouth spreads through the Lake District fell country, it will be the biggest disaster to hit the area during the 90 years of my lifetime – no, for even longer. Unless it is quickly checked, the disease will not only ruin the

---

35   The Rio Declaration stressed the need for a new approach to the general relationship between people and their environment. This tenet, accepted by governments and civil society alike, provided the context for the emergence of a consensus in the World Heritage Committee at its 16th session in 1992 which led to the creation of the new category of 'cultural landscape' (see above, note 4).

scenery of a million postcards, but shatter a long-established way of country life, wreck the tourist industry and destroy the area as we know it today.

This is because the fell farmers and their sheep – particularly the hardy, indigenous Herdwicks – are really the architects of the scenery. If the sheep are slaughtered and the fells cleared, it will change everything. Rough scrubland and coarse grasses will quickly take over the landscape. Moss, bogs and marshes will proliferate. Gorse and juniper will spread. The close-cropped upland pastures will disappear. The old stone walls will crumble. The whole landscape will harshen and deteriorate. The Lake District will disappear.[36]

The FMD outbreak also brought into sharp focus the interdependence of farmers and tourists. The Lake District was all but closed to visitors during the summer of 2001 and this not only provoked fury amongst walkers but caused hardship to farmers who relied on income from tourism. The knock-on effects, not least from publicity, persisted for several years. The FMD crisis could have been used to sharpen awareness of the need to manage the landscape in an interdisciplinary and holistic way, and galvanise policy-makers and rural managers and practitioners into action. It could have been the catalyst for developing a clearer understanding of the interrelationship between culture and nature, and between the environment and human use of that environment, by farmers and tourists alike. But the opportunity was not seized by many of the key players, even though the National Trust developed a Vision for the Lake District after Foot and Mouth that attempted to draw in all aspects of sustainability: economic, social, environmental and cultural.[37]

In 2007 the UK government ratified the European Landscape Convention (ELC), the first international convention to focus specifically on landscape. It defines landscape as 'an area, as perceived by people, whose character is the result of the action and interaction of natural and/or human factors'.[38] Initially, UK ratification was seen as a strong indicator that landscape would be put at the heart of government policy in a way that respected both nature and culture. Natural England, the body leading the implementation of the ELC, produced an Implementation and Action Plan. It encourages other organisations to do likewise. Nevertheless progress with raising the profile of landscape as the asset that frames protection and management has to a degree been overshadowed by other initiatives to address climate change in particular.

---

36 H. Griffin, 'If they go, it is the end of Lakeland', Country Diary, *The Guardian*, 11 April 2001.

37 *Lake District National Park Authority Promoting Understanding* (Audit Commission, January 2003), mentions the National Trust's Vision for the Lake District; the text is set out in S. Denyer, 'Valuing the Environment: the National Trust's role in protecting the cultural landscape of the Lake District', in R. Wylie (ed.) *Ethics to Expediency* (Westlakes Research Institute, 2002), pp. 54–5.

38 The European Landscape Convention (Florence: Council of Europe, 2000), Article 1.

**Figure 1.5   Hawkshead viewed across fields to the north of Esthwaite Water,
                 c.1910**
Source: © The author's collection.

In policy circles, the idea of landscapes as instruments of 'environmental service delivery' has gained a high profile since 2009. Increasingly, upland farmers are seen as upland managers who can help to deliver these landscape services, which include water absorption to stop flooding downstream, the storage of carbon, opportunities for recreation and the protection of rare and important species. But this approach can cause conflict between different players in the traditional and modern-sector management of the valleys. If the technically and visually elegant drainage of in-bye fields ceases to be maintained, in order to increase water absorption in the valley-head fields, the whole economic and social structure of the valleys built up over the past 800 years could unravel in much less than a generation. Once abandoned, the carefully engineered water management systems would be almost impossible to revive. Fields would revert to bog. On the northern edge of Esthwaite Water, the land between the lake and the villages of Hawkshead and Colthouse used to contain the most productive fields (and the site of important ploughing matches). Taken over by rushes which merge into alder scrub around the lake, they now form part of a protected Ramsar wetland site of international importance. The settlement of Hawkshead once stood proudly overlooking its lake. Now the view of the church and its surrounding houses from across the lake has vanished. The designated nature conservation area has been at the expense of the visual, social and cultural coherence of the cultural landscape (Figures 1.5 and 1.6). Around Hartsop, the fields at the northern end of Brotherswater are reverting

Figure 1.6   A similar view of Hawkshead today
The woodland is on former fields that have been fenced off for nature conservation and protected, according to Natural England, as 'an example of "hydroseral" succession from open water, through reed swamp, sedge fen, bog and carr woodland to drier oak woodland'. *Source*: © The author.

to bog as the streams cease to be cleared of boulders and gravel, despite appeals from the local community. And negative constraints are in place too, in respect of the myriad of stone walls that characterise many valleys. It is no longer possible to repair the walls with large cobble stones taken from the beds of becks, as they are protected for their fish and other wildlife.

Further potential threats are illustrated by the impact on grazing regimes and water courses from a project on water quality. The Lake District Still Waters Partnership was set up in 2002 as a coalition of organisations, including the National Trust, Natural England, Environment Agency and National Park Authority. It has reported that many of the lakes are polluted and have too much silt, and that this is having an adverse impact on fish such as Arctic Char. The remedies include controlling grazing regimes on the fells and trying to re-connect rivers to their floodplain by removing what are seen as 'canalised' embankments and thus allowing water to meander in valleys where sediment would then settle rather than building up in the lakes. Some of these flood plains are already designated as Special Areas of Conservation, which means that biodiversity conservation takes priority. But these floodplain areas are vital for farmers, and a crucial part of the cultural landscape. Ennerdale was partly afforested with conifers in the 1930s, one of the triggers for the agreement not to afforest the central lakes area.

In 2003 a Wild Ennerdale Partnership was formed with the stated intention of allowing the landscape to 'evolve naturally' by reducing human intervention. This has resulted in the local community being moved from a central role as shapers of the landscapes to onlookers or supporters of natural processes. Farmers have described how this change has led to a 'perceived loss of historical experience, cultural knowledge, and local identity'.[39] The rationale for wilderness within one of the most celebrated cultural landscapes of Europe, if not the world, is set out solely in terms of 'untamed nature' being a uniquely desirable goal.

While these examples demonstrate the wholly commendable shift towards management of larger areas at the scale of whole watersheds, the schemes are focused primarily on environmental conservation. Clearly it is essential for farming processes to work in harmony with the environment to achieve a sustainable use of resources, but sustainability is wider than environmental sustainability: it must respect cultural dimensions. Farming and other traditional uses of the landscape should not be optional extras. These examples also show the unanticipated impact of protective designations that can bring funding in their wake. There are a multitude of designations that protect the natural environment at both national and European level (and funding to support them) but there is still almost no protection for cultural aspects of the landscape, apart from specific features such as some buildings and archaeological sites. And there is still no formal recognition in the UK for the cultural processes that link people with their environment, nor funding to sustain them.

More importantly still, such issues raise fundamental questions on governance. In a world where accountability, participation and mutuality are properly counted as the basis of efficient decision making, one is bound to ask what vision for the Lake District is being followed by these projects and innovations, and whose values are being respected? Justification on the grounds that they foster change and keep the landscape alive is sometimes offered. A land manager in the Lake District said recently: 'We shouldn't be afraid of change. What we don't want is a landscape pickled in aspic or a museum landscape. We want a living landscape'.[40] Meandering rivers and bogs may be living in natural terms but they could spell be end of resilient communities and living, cultural landscapes.

Somehow the idea of natural beauty created by people's interaction with nature has slipped away or at least any notion that planned changes to the landscape need to be considered for their impact on its overall beauty, harmony or visual integrity. So too has consideration that changes might impact adversely on the collaborative social structures underpinning the agro-pastoral systems, even though in the past 20 years the idea of social capital[41] is now seen as a strong component for effective

39   I. Convery and T. Dutson, 'Rural communities and landscape change: a case study of Wild Ennerdale', *Journal of Rural and Community Development* 3, 1 (2008), p. 113

40   M. Rowe, 'Is this all set to change?', *Geographical Magazine*, October 2007.

41   R. Burton, L. Mansfield, G. Schwarz, K. Brown and I. Convery, *Social Capital in Hill Farming* (Penrith: International Centre for the Sustainable Uplands, 2005).

environmental management. Social organisations that involve trust, networks, common rules and connectedness between people can promote confidence in the management of resources facilitated by cooperation. And this is just the basis of the system of knowledge, skills, traditions and practice that already exists in the Lake District and has done so for at least eight centuries.

In third-millennium Britain, the agro-pastoral upland farming system is not robust. Many farmers have to supplement very small farming incomes with earnings from tourism and from grants and subsidies. As was said at an Environment, Food and Rural Affairs Committee into Farming in the Uplands in 2010, upland farmers are in trouble and there are few opportunities for agricultural diversification.[42] If the fell farms of the Lake Distrust are to survive, support for them will have to include non-market services for public goods; in other words payments that recognise the value of the landscape of which they are highly competent and conscientious stewards, and the benefits that that landscape delivers in cultural, social, environmental and economic terms. But what is that value and what are the benefits? Should the Lake District be supported in order to allow the valleys to revert to a state of wildness that has not been known for at least a thousand years or should the agro-pastoral cultural systems that have created a landscape perceived to be of great beauty and that have attracted and continue to attract missions of tourists each year be fostered and supported? These challenges were well summed up in an article in 2007: 'Someone has to decide whether we want the kind of views that we have here at the moment, which draw people in, or a view that looks more like parts of Scotland'. The current landscape is 'nice in itself but should it have a value placed on it? If the land is going to continue to be stewarded, what's it going to be stewarded to create?'[43]

It is estimated that as many as 15 million tourist-day-visits a year are made to the Lake District, naturally some in blocks of a weekend or a month. Together, they contribute some £6 million to the local economy. Most of the evidence suggests that tourists come primarily to see a cultural landscape stewarded by farmers. The head of Great Langdale, a valley whose settlement patterns date back to the twelfth century, is viewed by several million people each year but looked after by as few as six farmers. A castle of similar age visited by similar numbers of people and looked after by only six people would be deemed an extraordinary success story. Because there is no direct return from the Lake District tourism economy to its primary landscape managers, the farmers and farming families, and no directly measurable addition to human happiness, few economists or policy makers have considered the balance. Consequently, the landscape is not assessed in economic terms for the wealth and other benefits it generates. As the North West Uplands Farming Forum

---

42 *Farming in the Uplands*, House of Commons Environment, Food and Rural Affairs Committee: Examination of Witnesses: Dr Andrew Clark, Will Cockbain, William Worsley and Professor Allan Buckwell (10 November 2010).

43 Rowe, 'Is this all set to change?'

has advocated,[44] a well-researched formula of tourism value of each hectare of cultural landscape that could provide a benchmark for establishing a return income to farmers is urgently needed to close the loop between providers and users.

The potential World Heritage nomination for the Lake District provides the opportunity to address these issues and particularly how to define the value the Lake District landscape, and the attributes that convey that value. In 2005 a Lake District World Heritage Site Steering Group, chaired by Lord Clark of Windermere (a former Minister) and including representatives of local, county, regional and national organisations, was formed to develop a nomination for the Lake District and thus set out why the Lake District should be considered to have Outstanding Universal Value.[45] The Group considers that the Lake District landscape is exceptional for the way it has been shaped by long-standing agro-pastoral traditions, is perceived to be a great beauty and harmony, has been celebrated by writers and artists, reflects picturesque ideals, and has become a testimony to the foundation and successes of the conservation movement.

The Lake District landscape is thus valued not as a natural landscape but as a cultural landscape, an exceptional cultural landscape that might be seen to have global value. There is some evidence that this value is now being recognised by the Lake District National Park Authority. In October 2011, in response to criticism that it did not value the contribution of farmers in the maintenance of the Lake District landscape, and was 'was not on their side', its Strategy and Vision Committee adopted a Commitment to Working with Farmers.[46] This includes policies to 'Connect the special qualities of the National Park, including landscape features and cultural heritage, to the economy through imaginative approaches, including use of traditional skills' and 'Actively support land managers in the task of sustainably managing the landscape: delivering environmental and economic benefits for themselves and the wider community'. More crucially it acknowledges some of the reasons behind the proposed World Heritage nomination in terms of the resilience and persistence of the farming culture which 'sets the Lake District apart from many other national and international regions where this has been lost or inappropriately compromised'. Perhaps it is now being recognised that environmental concerns have been given far too much priority over agriculture and the cultural traditions on which it is based. This document is a major step forward. It recognises that the physical and cultural landscape created by farmers is a major draw for the 15.8 million visitors that come to the Lake District each year. And it sets out the need for future activities to increase their focus on 'Community-led Valley Planning'.

---

44   *Farming in the Uplands*, House of Commons Environment, Food and Rural Affairs Committee: Written evidence submitted by North West Upland Farming Forum (9 February 2011).

45   Lake District World Heritage project web-site: http://www.lakeswhs.co.uk/.

46   *Our Working Relationship with Farmers* (Lake District National Park Authority, Park Strategy and Vision Committee, 26 October 2011).

Great stress is now put by the UNESCO World Heritage Committee on the need for community participation in World Heritage properties as part of collaborative governance arrangements that draw in key stakeholders. The nomination of the Lake District for World Heritage status could be the mechanism that allows stakeholders to define a common vision for the Lake District, that sets out how this dynamic, resilient landscape might be sustained so that its traditional processes and collaborative associations are respected as part of the social and human capital accumulated within the farming systems, and how the valleys will continue to convey their beauty and harmony and deliver their 'wealth' in the widest meaning of the word. The Brundtland Report noted that responsibilities for decision making are being taken away from both groups and individuals. Perhaps the World Heritage bid is the opportunity to put local communities back at the heart of the decision-making process?

The Lake District is one of the last examples of agro-pastoral farming cultures in the UK and we stand to lose it unless there is a clearer understanding of what it is, its value, the benefits it delivers and how it might be sustained in a dynamic, collaborative way. The landscape has survived so far as a result of an extraordinary fusion between those who work the land, and those who have adopted it as their home and had the foresight and influence to ensure its protection. Visitors for the last 250 years have been part of this symbiotic relationship contributing to its prosperity: their support is needed for the future.

Chapter 2
Setting the Scene

John K. Walton

The English Lake District is widely agreed to be an outstandingly beautiful area, offering literary associations and adventurous play as well as landscape to the discerning tourist. It presents itself not only as a National Park (on the distinctive and limited British definition) but also as a candidate for UNESCO World Heritage Site status, for which it would be required to demonstrate 'Outstanding Universal Value'. The actual distribution of such sites is strongly skewed towards the wealthier, more developed and more politically influential parts of the globe, as Frey, Pamini and Steiner have explained, with 46 per cent of the total list being located in Europe. But their econometric analysis does not extend to examining the bias in cultural preference that seems also to be evident in the compilation of the list.[1] Although the Lake District has yet to feature on the World Heritage list, it has long been deemed a plausible contender, held back mainly by categorisation problems; and the cultural preference point is illustrated by the fact that its landscapes do not have universal appeal across the globe, or even within Britain.[2] Nor has their appeal been universal across time: until as recently as the mid-eighteenth century travel and topographical writers such as Celia Fiennes and Daniel Defoe were repelled and even frightened by these howling wastes and barren wildernesses, wanting to feast their eyes on lush meadows and productive fields, on tidy, managed, controlled, reassuring tracts of agricultural lowlands, with the human hand visibly in command of a tamed and subdued landscape.[3]

---

1   B.S. Frey, P. Pamini and L. Steiner, *What Determines the World Heritage List? An Econometric Analysis* (University of Zurich, Department of Economics, Working Paper No. 1, January 2011). For the origins of the World Heritage concept, M. Hall (ed.), *Towards World Heritage: International Origins of the Preservation Movement, 1870–1930* (Farnham: Ashgate, 2011).

2   Recent research suggests, however, that even John Constable, strongly identified with the very different landscapes of rural Suffolk, was not as daunted and dismayed by the Lake District as was once thought, but responded with creative enthusiasm on an extended visit in 1806: S. Hebron, C. Shields and T. Wilcox, *Constable and the English Lake District* (Grasmere: Wordsworth Trust, 2006).

3   N. Nicholson, *The Lakers: the Adventures of the First Tourists* (London: Robert Hale, 1955). I. Thompson, *The English Lakes: a History* (London: Bloomsbury, 2010), pays homage to Nicholson but offers no real advance on his work: the text is slight and under-researched, although the photographs are excellent.

The perceived attractiveness of Lake District landscapes in Western cultures is the product of a revolution in cultural values during (mainly) the second half of the eighteenth century, which transformed perceptions of the aesthetics of coastal as well as mountain scenery and associations, making maritime and upland landscapes attractive, evocative and exciting under the rubrics of the Picturesque and then the sublime, and in response to intimations of God's presence in Nature.[4] These changing perceptions in cultivated society, becoming visible at mid-century, made the Lake District fashionable from the 1760s and 1770s onwards as its scenery was marketed by entrepreneurs of taste and transport improvements began to make it more accessible. We still lack a thoroughgoing investigation of how fast, how far and to what extent the new values trickled down through the middle ranks and eventually affected the working class, but this clearly involved a complex and variegated set of processes. It is clear that not everyone has fallen under the spell of the Lake District's magic. It is mainly an Anglophone, even an English, attraction: a very thorough recent French study of the timing and geographical pattern of the development of 'lieux touristiques' across the globe since the eighteenth century makes no mention of the English Lake District, and if it forms part of the heavily overlapping cluster of early sites in England the scale of the global maps presented does not allow its identification.[5] Even among visitors to the area the conventional attributes of beauty, tranquillity and space for contemplation (whether the watcher is in repose or in pedestrian motion) can be trumped by the opportunities for adventure and excitement, sometimes involving disruptive or physically damaging technologies of movement, from off-road motor vehicles to power-boats and powered hang-gliders, as we shall see. But rapid assimilation of the new values in aspiring sectors of the emergent commercial middle classes of the later eighteenth century is indicated by the behaviour of the Langton family, linen merchants of the small town of Kirkham in west Lancashire, in the mid-1770s. They not only identified and followed the new fashion for making a picturesque tour of the Lakes: they kept detailed travel diaries with an opportunistic eye to the income-generating potential of a possible publication, reminding us of an instrumental and exploitative dimension to landscape enjoyment, which is capable of coexisting with admiration and appreciation and has been of enduring importance.[6]

The eighteenth-century transformation of attitudes to landscape and scenery in Western Europe was absent in many other societies, as one of my doctoral students found during the 1980s when introducing the Lake District to parties of visiting Malaysians, who found the upland landscapes barren, ugly and unappealing, and could not understand why they were being invited to admire them. The low

---

4   A. Corbin, *The Lure of the Sea* (Cambridge: Polity, 1994); M.H. Nicolson, *Mountain Gloom and Mountain Glory* (Ithaca: Cornell University Press, 1959).

5   Équipe M.I.T., *Tourismes 3: La Révolution Durable* (Paris: Éditions Belin, 2011), pp. 125–8.

6   J. Wilkinson, *The Letters of Thomas Langton, Flax Merchant of Kirkham, 1771–1788* (Manchester: Chetham Society, 1994).

incidence of British Asians among visitors to the Lake District (1 per cent of visitors compared with 9 per cent of the relevant population, on one recent calculation) has prompted efforts to redress the balance by alerting this constituency to the delights of Lakeland landscape and environment, as well as reassuring them that they will not be conspicuous or out of place.[7] Recent research has highlighted a continuing lack of affinity for this kind of landscape among most British Asians, unless they have encountered it in positive ways through school, university or social contact. It also underlines the ways in which its appreciation may be viewed as a native 'English', or at least Anglophone, cultural possession which is not expected to be shared.[8] Such inhibitions are, of course, not confined to British Asians. Significantly, the boom in Japanese visitors to the Lake District in the early twenty-first century, leading to a proliferation of Japanese pictograms at Oxenholme railway station and the Bowness boat landings on Windermere, has had nothing directly to do with landscape appreciation: it was based on the popularity in Japan of the Beatrix Potter 'Peter Rabbit' books, which were translated in 1971 and became much-loved school texts for the learning of basic English. The place of pilgrimage was Potter's home, Hill Top Farm at Sawrey: a replica of the building was constructed at a children's zoo in Tokyo, and at about the same time the film 'Miss Potter', launched in 2006 with Renée Zellweger in the title role, created a further spike of interest. This was the main centre of attraction for Japanese visitors, although they are also finding their way to Dove Cottage and other Lake District 'pilgrimage' sites in growing numbers, as well as to Haworth in pursuit of the Brontë sisters.[9] We should not overlook the Chinese academic Chiang Yee, however, whose drawings and verses inspired by Lake District scenery in 1936 suggested that here was also a culture that might embrace the Lake District, especially when we set this alongside the evidence of Chinese interest in Wordsworth, Burke and the sublime in a context of pantheistic mysticism. Yi-Fu Tuan's work reveals a very long tradition of Chinese appreciation of mountain landscapes, which became visible in eighteenth-century England through imported pottery and artwork, much of which was, however, assimilated into an 'aesthetic of exoticism'.[10]

---

7   K. Baden, 'Asians and the Great Outdoors', *Asian Image* (7 Dec. 2009), http://www.asianimage.co.uk/profiles/471890/Asians_and_the_great_outdoors/ (accessed 30 March 2011).

8   D.P. Tolia-Kelly, 'Fear in Paradise: the affective registers of the English Lake District landscape re-visited', *The Senses and Society* 2 (2007), pp. 329–51.

9   http://www.news.bbc.co.uk/1/W/england/cumbria/3070620.stm (accessed 30 March 2011). They may have been startled by the recent installation of Mr. McGregor as Director of the Wordsworth Trust.

10   C. Yee, *The Silent Traveller: a Chinese Artist in Lakeland* (London: Country Life, 1937). This book was the first in a series that eventually covered several British locations. See also Y. Zheng, *From Burke and Wordsworth to the Modern Sublime in Chinese Literature* (West Lafayette, Indiana: Purdue University Press, 2011); Y.-F. Tuan, *Space and Place* (London: Edward Arnold, 1977), p. 57; D. Porter, *The Chinese Taste in Eighteenth-Century England* (Cambridge: Cambridge University Press, 2010).

The further dimension that marked out the Lake District as a sacralised Western (and particularly English) literary landscape, was its association with the work of William Wordsworth and the 'Lake Poets', including Thomas De Quincey, probably the only editor of the *Westmorland Gazette* to have been addicted to opium.[11] Keith Hanley's chapter develops the Wordsworth theme; but the work of Jonathan Bate, tracing a genealogy of perception and influence from Wordsworth to the formal foundation of the National Park in 1951, has been particularly influential here, making links to 'eco-criticism' and 'green socialism'.[12] But the 'Lake Poets' dimension has also excluded as well as included, in that full appreciation requires a cultural repertoire which is distinctively English (or Anglophone or anglophile, as Melanie Hall's chapter demonstrates), although the list of well-known and influential authors connected with the area is much longer and more diverse than an exclusive focus on the Romanticism of the late eighteenth and early nineteenth century might suggest. Wordsworth remains at the core, with Dove Cottage (especially) and Rydal Mount prominent among the Lake District's most popular 'visitor attractions', although it should be emphasised that Dove Cottage would come only fifteenth in Cumbria Tourism's list of the top twenty attractions in the county for 2009 (even if it had submitted its figures), and eighth among attractions within the Lake District National Park. Here, the most popular entries by far were the lake cruises on Windermere and Ullswater, followed by the Forestry Commission forest parks at Grizedale and Whinlatter, the visitor centre at Brockhole, Beatrix Potter's Hill Top Farm, the Cumberland Pencil Museum and the Beatrix Potter Gallery.[13] Moreover, Wordsworth's contemporary 'Lake Poet' Robert Southey, a literary lion of his time who spent much of his adult life among the Lakes, has receded from the non-specialist gaze, despite recent attempts at rehabilitation which are more concerned with other aspects of his life and work.[14] This should remind us, at the outset, that the academic and 'high culture' criteria for evaluating the Lake District as a special environment are not always, or even usually, in tune with popular preferences, which incline towards comfortable enjoyment of picturesque views from lake level (Wordsworth was on record as opposing the first steamers on Windermere), the use of Forestry Commission conifer plantations as adventure playgrounds, encounters with anthropomorphic animals, driving around, picnics, urban or village sightseeing, shopping, art galleries and a little light industrial archaeology.[15] A survey in 2008 found that the two most popular destinations for coach tours within North-West

---

11   R. Caseby, *The Opium-Eating Editor* (Kendal: Westmorland Gazette, 1985).

12   J. Bate, *Romantic Ecology: Wordsworth and the Environmental Tradition* (London: Routledge, 1991).

13   http://www.cumbriatourism.org/research/attractions.aspx (accessed 7 April 2011).

14   W.A. Speck, *Robert Southey: Entire Man of Letters* (New Haven: Yale University Press, 2006).

15   P.A. Blakey, D.W.G. Hind and C.A. Rawding, 'The Lake District as a tourist destination in the 21st century', in D.W.G. Hind and J.P. Mitchell (eds), *Sustainable Tourism in the English Lake District* (Sunderland: Business Education Publishers, 2004), chapter 4.

England were Blackpool (50 per cent) and the Lake District (30 per cent); and it is probable that many of the (mainly retired) customers would happily go to either destination, and (in defiance of late Victorian expectations) might even seek similar pleasures in each of them.[16] We should also remember, however, that the individual destination statistics are for locations that levy an admission charge, and do not take account of the free attractions of Wordsworthian pedestrianism and contemplation on the lake shores and open fells.[17]

Wordsworth himself has, indeed, dominated the association of the Lake District with the Romantic Movement, with outposts of pilgrimage at Hawkshead and Cockermouth, almost at opposite ends of the National Park, supplementing the main attractions at Grasmere and Rydal. His coevals have faded from popular visibility, and at a popular level Wordsworth himself is overwhelmingly associated with a single poem, the sonnet that begins 'I wandered lonely as a cloud', better known as 'Daffodils', which is firmly located (via his sister Dorothy's journal) on the north shore of Ullswater in the heart of the National Park, and continues to feature on school syllabi in Britain and across the former British Empire, while providing a fecund stimulus for irreverent parody.[18] Wordsworth's Duddon sonnets, celebrating a less accessible river on the Lake District's south-western fringe (as discussed later in this volume by David Cooper), or his poems in praise of the celandine, have attracted much less popular recognition or interest. His *Guide to the Lakes* (usually thus abbreviated), which has remained in print for over two centuries as an essential *vade mecum* for serious tourists, has been presented as a pioneering exemplar of holistic regional geography; and his general visibility remains very high.[19] Subsequent Lakeland literary lions have been less enduring, although the visionary art critic and controversial social philosopher John Ruskin, whose home at Brantwood on Coniston Water became a place of pilgrimage in the first half of the twentieth century, is now enjoying something of a revival.[20]

Another controversial figure, of much lesser stature but with a reputation in several fields, was Arthur Ransome, who emulated Beatrix Potter by writing children's stories (but with the children as imaginative protagonists), in the *Swallows and Amazons* series, which could be imaginatively located at various points on and around Windermere and Coniston Water, and attracted enduring

---

16   S.J. Page, *Tourism Management: Managing for Change* (London: Butterworth Heinemann, 2009), p. 188.

17   J.K. Walton, 'The Windermere tourist trade in the age of the railway, 1847–1912', in O.M. Westall (ed.), *Windermere in the Nineteenth Century* (Lancaster: Centre for North-West Regional Studies, 1976), pp. 19–38.

18   http://www.wordsworth.org.uk/poetry/index.asp/paged=101 (accessed 7 April 2011).

19   W. Wordsworth, *A Guide through the Scenery of the Lakes in the North of England* (Kendal: Hudson and Nicholson, 1835); I. Whyte, 'William Wordsworth's *Guide to the Lakes* and the geographical tradition', *Area* 32 (2000), pp. 101–6.

20   K. Hanley and J.K. Walton, *Constructing Cultural Tourism: John Ruskin and the Tourist Gaze* (Bristol: Channel View, 2010), chapter 7.

devotion from children alongside the attempted topographical reconstitution of sites and journeys by obsessive adults.[21] Karen Welberry has argued that Ransome should be given credit for providing an alternative justification for the special nature of the Lake District, with a focus on adventure and the exercise of the imagination which incorporated the urban and industrial aspects of landscape and society, and was accommodating to change. She traces Ransome's appreciative genealogy from Ruskin through W.G. Collingwood, whom she regards as an undervalued figure in the dominant discourse, working along the alternative lines of footpath preservation and Ambleside's Armitt Library rather than the National Trust (to which we shall return) or the 'romanticists'; and she points to the absence of any Wordsworthian references in Ransome's *Swallows and Amazons* books, while suggesting that Ransome has been marginalised because of low academic esteem for children's literature. This is an interesting attempt at reappraisal, but it falls far short of establishing any but the most speculative links between Ransome's work and the foundation of the National Park, a relationship which is founded on mere assertion.[22] It would, indeed, be hard to base acclaim for 'Outstanding Universal Value' on Ransome's output, and the same applies to authors with local connections such as Hugh Walpole and Melvyn Bragg, despite the popularity for a generation of the former's 'Herries Chronicles' trilogy of the early 1930s and the (limited) Lake District output of the latter, even allowing for his enduring prominence in Britain as a cultural entrepreneur.[23] In both cases the Lake District output is part of a much broader range of publications, and lacks the international visibility coupled with a powerful emotive connection to a distinctive vision of native soil and society, bringing the local on to a global stage, that Wordsworth provides.

It was indeed and above all the intimate Wordsworth connection, which extended, as we shall see, to claims for a unique relationship between poetry, locality, topography, landscape and a virtuous society, which enabled the Lake District to stake a claim to 'Outstanding Universal Value' as a potential UNESCO World Heritage Site, despite a lack of genuine universality which was shared by many, and perhaps most, of the already-designated locations. As had previously become apparent, the extent to which its landscapes had been shaped and moulded by human intervention, especially over the last millennium (and as discussed

---

21   J. Sparks, *Arthur Ransome's Lake District* (Wellington: Halsgrave, 2007) is the most recent, but see especially various works by R. Wardale, including his *Arthur Ransome and the World of the Swallows and Amazons* (Skipton: Great Northern Books, 2000).

22   K. Welberry, 'Arthur Ransome and the Conservation of the English Lakes', in S.L. Dobrin and K.B. Kidd (eds), *Wild Things: Children's Culture and Ecocriticism* (Detroit: Wayne State University Press, 2004), chapter 5. It is surprising that Welberry does not make more of the location of the original meeting between Ransome and Collingwood on which she lays great emphasis, in the former industrial location of Coniston's Coppermines Valley.

23   R. Hart-Davis, *Hugh Walpole* (London: Macmillan, 1952); M. Bragg, *The Maid of Buttermere* (London: Hodder and Stoughton, 1987).

below in Chapters 3 and 4), debarred it from consideration as a purely 'natural' phenomenon.[24] Like most English landscapes, those identified with the Lake District are palimpsests, embodying layer upon layer of past agricultural practices, transport systems and extractive and manufacturing industries, and sustained by generations of human migration into, out of and within the area. The industrial past of the National Park includes environmentally intrusive traces of mining for copper, lead and other ores, with related washing and smelting activities, and the enduring scars of slate quarrying.[25] Industrial archaeology and museums now provide distinctive visitor attractions alongside and in conjunction with literary landscapes, adventure tourism and souvenir shopping.[26] Railways penetrated tentatively and unobtrusively into the periphery of the future National Park: the only trunk route to pass through the district was the Cockermouth, Keswick and Penrith line, which was more important for freight traffic serving West Cumberland industries than for tourists, and survived for just over a century after its opening in 1864.[27] The widened dual-carriageway road which effectively replaced it in the mid-1970s was much more intrusive, although a contemporary survey found local reactions to be generally positive. Problems of through heavy goods traffic increasingly compounded the pressures of rising demand from motorised tourism, including the expansion of car parking space, throughout the second half of the twentieth century, and beyond.[28]

The demand for water from external industries and populations brought reservoir development into the heart of Lakeland at Thirlmere as well as (half a century later) on its eastern fringe at Haweswater (both promoted by city government from as far away as Manchester, and entailing extensive earthworks and engineering, and the submersion of villages), and, much less intrusively, in its north-western corner at Crummock Water (Workington) and Ennerdale (Whitehaven). Several other projects never came to fruition, not least because of the fierce opposition they provoked (as with late Victorian railway proposals in sensitive areas, although the dubious economic viability of these was more of an issue); but the first of the Manchester interventions provoked the fiercest opposition from defenders of the 'natural' landscape of the Lakes from the

---

24 See above, Chapter 1. This is not to endorse the absurd anthropocentric assumption that 'nature' is a linguistic construction, and has no existence outside human discourse and perception: see E. Crist, 'Against the social construction of nature and wilderness', *Environmental Ethics* 26 (2004), pp. 4–24.

25 J.D. Marshall and J.K. Walton, *The Lake Counties from 1830 to the Mid-Twentieth Century* (Manchester University Press, 1981).

26 J.D. Marshall and M. Davies-Shiel, *The Industrial Archaeology of the Lake Counties* (Beckermet: Michael Moon, 1977); E. Holland, *Coniston Copper* (Milnthorpe: Cicerone Press, 1987).

27 D. Joy, *The Lake Counties: the Regional History of the Railways of Great Britain, Vol. 14* (Newton Abbot: David and Charles, 1983).

28 P. Prescott-Clarke, *People and Roads in the Lake District: a Study of the A66 Road Improvement Scheme* (Crowthorne: Transport and Road Research Laboratory, 1980).

late 1870s onwards, and helped to put the Lake District on a global academic agenda as the new discipline of environmental history emerged at the turn of the millennium.[29] The Forestry Commission, established in 1919, brought another kind of industry into the Lake District; the systematic, managed production of timber to sustain a strategic reserve in response to shortages during the First World War, resulting in extensive mono-cultural geometrical pine forests which were expanded rapidly on poor upland soils during the 1920s and 1930s, acquiring deciduous outer fringes in places in response to criticism. The undesirable spread of 'artificial plantations' of conifers, in the preferred contemporary form of the larch, had been a preoccupation of Wordsworth more than a century earlier.[30] The Commission's dominant product at this time was pit props, literally to shore up the workings of the economically indispensable coal industry, and its expansionist activities generated fierce resistance from those who valued the landscapes on which the new conifer forests encroached, which were themselves the products of earlier phases of human activity on the hills. The limited coverage of Lake District issues in the Commission's official history suggests that this was an asymmetrical relationship, but the Commission's activities were a major issue for the defenders of Lakeland scenery and ambience.[31] Arthur Gardner, writing in 1942, listed all these threats to the integrity of the Lake District, and added 'buildings in the wrong places and unsuitable designs', diverging from the local vernacular and appropriate scale; 'power schemes, pylons and overhead wires', which brought the benefits of electricity relatively cheaply but threatened the pristine simplicity of the landscape (the underground alternative was adopted in practice within the Lake District proper, but only after strong and sustained pressure from influential lobbyists from the conservation societies); and the ubiquitous problem of 'vulgarisation by the spreading of litter', which could, he thought, only be overcome by improved education.[32]

Looking outwards from the Lake District, many adjacent landscapes which are visible from or within reach of the fell tops offer less than pristine prospects, as David Cooper reminds us with special reference to Millom, and Jason Wood demonstrates in his examination of the complex relationships between landscape, heritage and industry that inflected the role of Furness Abbey as a 'gateway to

---

29   Marshall and Walton, *Lake Counties*, chapter 9 and conclusion; J. Winter, *Secure from Rash Assault* (Berkeley: University of California Press, 1999); H. Ritvo, *The Dawn of Green: Manchester, Thirlmere and Modern Environmentalism* (Chicago: University of Chicago Press, 2009).

30   Wordsworth, *Guide*, pp. 162–8.

31   H.H. Symonds, *Afforestation in the Lake District* (London: Dent, 1936); I.O. Brodie, *Forestry in the Lake District* (Kendal: Friends of the Lake District, 2004), which takes a 'retrospective' look at Symonds' book; G. Ryle, *Forest Service* (New York: Augustus M. Kelley, 1969), p. 259.

32   A. Gardner, *Britain's Mountain Heritage and its Preservation as National Parks* (London: B.T. Batsford, 1942), pp. 19–25. For pylons and reactions to them, B. Luckin, *Questions of Power* (Manchester University Press, 1990).

Lakeland' The views westwards from the central mountain dome incorporate a coastal industrial fringe, from Barrow to Maryport, containing the remnants of Victorian 'carboniferous capitalism' in coal and iron mining, steelworks and shipbuilding. More obvious now are the results of various attempts to alleviate the consequences of the inter-war depression (which had particularly deep roots in this area, as seams and pits were already being worked out before the First World War), including a variety of wartime and post-war initiatives, especially the Marchon chemical plant on an old colliery site at Whitehaven.[33] Most intrusive of all were the successive arrivals of the state-sponsored nuclear weapons, power and reprocessing industries at Windscale (where most of the site reverted to the original name of Sellafield for cosmetic reasons in 1988), beginning in the immediate aftermath of the Second World War, and already under development when the Lake District National Park was designated in 1951. This could still be seen in the early 1960s as a positive emblem of progress and modernity, and it certainly worked well as a job creation scheme in an area of persistent high unemployment, which the Forestry Commission was also trying to tackle.[34] But it is also widely regarded as a particularly unfortunate, even philosophically inappropriate, intrusion into the Lake District's immediate surroundings, as the Millom poet Norman Nicholson reminded his readers in 1972, when he made a direct link between the 'toadstool towers' of Windscale, the view from Scafell and the invasion of the Lake District by the cancerous emanations of an unconscionable industry, founded on the production of weapons of mass destruction.[35] The fall-out from the near-catastrophic Windscale accident of 1957 penetrated the National Park, and these and other emissions may have enduringly contaminated the western fells in ways that only became apparent when tests were made in the aftermath of the Chernobyl accident nearly thirty years later.[36] Meanwhile, in the prospect northwards from the back of Skiddaw the four 300-foot cooling towers of the Chapelcross nuclear facility near Annan, whose main purpose was to produce plutonium and later tritium for the nuclear weapons programme, also constituted a prominent landscape feature between opening in 1959 and demolition in 2007.[37] Subsequently the nuclear power complex at Heysham, which began to operate in 1983, inserted itself prominently into southward views from the heart

---

33   G.J.H. Daysh and E.M. Watson, *Cumberland, with Special Reference to the West Cumberland Development Area: a Survey of Industrial Facilities* (Whitehaven: Cumberland Development Corporation, 1951); F.J. Monkhouse, 'Cumbria', in J.B. Mitchell (ed.), *Great Britain: Geographical Essays* (Cambridge: Cambridge University Press, 1962), pp. 429–30.

34   Monkhouse, 'Cumbria', p. 430.

35   N. Nicholson, *A Local Habitation* (London: Faber, 1972). See also K. Hanley, 'The discourse of natural beauty', in M. Wheeler (ed.), *Ruskin and Environment* (Manchester: Manchester University Press, 1995), p. 37.

36   B. Wynne, 'Misunderstood misunderstandings: social identities and public uptake of science', *Public Understanding of Science* 1 (1992), pp. 281–304.

37   http://www.banthebomb.org/archives/scotland/chaacc.html (accessed 7 April 2011).

of the National Park.[38] Nicholson's dismay echoed the rhetorical despair of John Ruskin, nearly a century earlier, at the very thought of a single smoking chimney from Barrow's industries intruding into the skyline at his Brantwood sanctuary, although in contrast with Windscale's plutonium and other infernal elements his Victorian 'storm clouds' of carboniferous capitalism were at least both visible and, in themselves, of finite duration on a human scale.[39]

The visibility of the nuclear industry from within the National Park, whose western border comes within a couple of miles of Windscale at Calder Bridge, highlights an alternative or complementary vision of landscape heritage, which celebrates landmarks in technological innovation, and seeks to preserve them and make them available as visitor destinations, as industrial archaeology. There was local opposition to the demolition of the Chapelcross cooling towers, as local landmarks and 'proud relics of an industrial past', and during 2006–7 proposals were prepared to develop Calder Hall, the world's first nuclear power station to be connected to a national electrical power supply network, as a heritage tourism attraction. Calder Hall, which operated between 1956 and 2003, was part of the Windscale/Sellafield nuclear cluster, and a feasibility study pointed out the positive importance of its proximity to the Lake District, with its enormous visitor numbers and economic dependence on tourism, and its potential educational role for schoolchildren and others. Sellafield had its own visitor centre between 1988 and 2008, which underwent several metamorphoses before being replaced by a conference facility as its popularity declined.[40] Using this alternative scale of values, a case has indeed been made for submitting Calder Hall itself as a potential World Heritage Site, although problems of access, safety, security and deterioration of fabric would present challenges to the further pursuit of any such proposal.[41] Identification with this agenda need not, of course, preclude a strong parallel attachment to, and desire to preserve and protect, the landscapes of the National Park itself.

Those who worship at the shrine of the Lake District as a unique distillation of picturesque and sublime natural beauty do so overwhelmingly from an Anglo-Saxon perspective, and often from a patriotic English standpoint (which might complicate perceptions of Windscale), although Melanie Hall's chapter brings out the importance of the Anglo-American and Canadian dimensions. Many

---

38　For context see W. Patterson, *Going Critical: an Unofficial History of British Nuclear Power* (London: Paladin, 1985).

39　M. Wheeler, 'Introduction', in Wheeler (ed.), *Ruskin and Environment*, pp. 1–2. Wheeler chose to compare Barrow's smokestacks with the visual impact of wind farms, rather than the nuclear industry.

40　http://www.nda.gov.uk/documents/upload/NDA-Calder.Hall-Nuclear-Power-Station-Feasibility-Study-2007.pdf (accessed 7 April 2011).

41　English Heritage, *England's World Heritage Sites: Report to the Secretary of State. Tentative List Review Committee, Annex D: Sites Considered by the Committee* (London: English Heritage, 1998), p. 31.

English landscapes have been, and are, represented as essential emblems of national identity, of course, and the Lake District is distinguished from most of them by its lack of nucleated villages clustered around churches and manor houses, timber-framed cottages or 'patchwork quilts' of arable fields, as enclosed (usually) in the eighteenth century. Its landscape and architecture are distinctive, as Wordsworth noted.[42] Gardner, a particularly conservative commentator writing during the Second World War, began his paean to 'Britain's mountain heritage' with a seemingly inappropriate evocation of just this sort of landscape, 'lived-in' by successive generations and looked after and enhanced by the great landed families. Even so, he then achieved a seamless adjustment to incorporating the Lake District as part of this picture and pattern, despite the historic near-absence of resident large landowners or their conventional impact on the landscape, while emphasising its small scale, manageability and vulnerability in ways that have been shared by many advocates of National Parks as distillations of imagined English, or British, political as well as aesthetic virtues.[43]

Malcolm Andrews has examined the resolution of an apparent paradox whereby ideas about the positive relationship between 'Picturesque' landscapes and the virtues of a simple rural life were taken from Italian painters (Claude, Salvator Rosa) and classical authors (Horace, Virgil), given an 'English vernacular flavour', and pressed into the service of emergent eighteenth-century English (or British) national pride. As English vernacular authors such as Milton and Gray displaced the classical originals, the definition of the 'man of taste' became less culturally (or at least linguistically) demanding, and, in the Lake District, the new commercial and industrial wealth of Lancashire and West Yorkshire (as well as the metropolis) craved 'prospects' and embellishments, as Adam Menuge discusses later in this volume. The pattern of country house and villa development in the Lake District, from the late eighteenth century onwards, with clusters of big houses in parkland and garden settings concentrated especially around the accessible centres of Windermere, Troutbeck, Ambleside, Keswick and Coniston, adds a distinctive cultivated dimension to lakeside landscapes which does not necessarily sit easily with a romantic aesthetic of wild beauty and rustic simplicity, although such buildings often pay homage to local vernacular, which is itself a complex concept. But the employment of high-profile, sometimes innovative architects and garden designers (such as Voysey, Baillie Scott and Mawson) produced distinctive and attractive cultural landscapes of the cultivated kind, which attracted recognition in their own right as contributions to a broader picture.[44]

---

42 D. Matless, *Landscape and Englishness* (London: Reaktion, 1998).

43 Gardner, *Britain's Mountain Heritage*, pp. 1, 5–6.

44 M. Andrews, *The Search for the Picturesque: Landscape Aesthetics and Tourism in Britain* (Stanford University Press, 1989), pp. viii, 3–4, 6–7, 9, 12; M. Hyde and N. Pevsner, *The Buildings of England: Cumbria* (New Haven and London: Yale University Press, 2010), pp. 67–70, 101–3, 169–76, 295–7, 377, 450–2, 642–4, 691–4.

Despite apparent tensions and contradictions, the generally unproblematic perception of the place of a conventional vision of the Lake District in British (especially English) culture is indicated by its limited representation in the British Cartoon Archive, where already in 1904 a cartoon by W.K. Haselden, lamenting the annual tourist exodus to 'the Continent', placed it at the top of a list of 'Beautiful Places in Great Britain'. A rare appearance by the Lakes in 1946, when post-war sites for military training were under discussion, made its joke by portraying the laughably unthinkable (as well as impractical), as a caricature Blimpish officer instructed a subordinate, 'And you might get the sappers to fill in the lakes we won't be requiring'. During the invented crisis over foot and mouth disease in 2001, which is discussed in Chapters 1 and 4, a *Times* cartoonist drew a stylised depiction of the Lakes as pools of blood from slaughtered sheep, again drawing currency from an external threat to a generally inviolable national sanctuary. Such perceptions could be traced back at least to the late nineteenth century, when Ruskin was caricatured in 1876 as a defender of the Lake District against 'railway vandalism', while in 1883 a campaign against a proposed mineral railway in a sensitive location stimulated a cartoon representing the intruder as the 'steam dragon of Honister', which had to be slain by St George in defence of the ladies of the Lakes.[45]

Such processes have both responded to and furthered the development of perceptions such as Gardner's which identify the Lake District with specifically British (or indeed English) cultural sensibilities, and thereby make it seem less accessible to those from beyond the charmed circle. Current discussions of 'multiculturalism' and its discontents make such identifications particularly contentious, especially (for example) in the light of the apparent (self-) exclusion of many (though certainly not all) British Asians from such (cultural) landscapes. An exclusive 'multi-culturalism' which advocates the keeping of cultures in separate, impermeable compartments would leave the Lake District as a symbolic redoubt of Englishness, or at least of international Anglophone culture, and thereby imperil the whole notion of 'Outstanding Universal Value', unless its universality can be recognised from without by those who do not actually embrace its values for themselves. But the alternative version, 'multicultural-ism', which embraces diversity, dialogue, interaction and even mutual borrowing, opens out the potential for universality in settings like the Lake District, while undermining such claims if that potential remains unrealised.[46] Can appreciations of the Lake District pull together, and universalise, the local, the national, the Anglophone trans-national and the global?

---

45   British Cartoon Archive, University of Kent, WH 0025 (W.K. Haselden, 1 July 1904); NEB 0464 (R. Nebour, 27 August 1946); 57205 (P. Brookes, *Times*, 20 March 2001); *Punch* (5 February 1876); Marshall and Walton, *Lake Counties*, chapter 9; and Jason Wood in chapter 11, below.

46   M. Hasan, 'How we rub along together', *New Statesman* (4 April 2011), p. 30; B. Paresh, *Rethinking Multiculturalism* (Cambridge, Mass.: Harvard University Press, 2002).

One route towards such an outcome might lie through the role of the 'defence' of the Lake District in the nineteenth-century emergence of environmentalism, as indicated above.[47] Wordsworth himself was a strong advocate of preserving the Lakeland environment, in the broadest sense. From 1810 onwards, and especially after the most influential edition appeared in 1835, his *Guide to the Lakes* provided a regularly reprinted and widely read showcase for his opinions, from architecture to gardens and tree planting, but always taking full account of the social systems that sustained and were inseparable from the maintenance of the desired landscape and scenery.[48] He was, of course, an early campaigner against the extension of railways beyond the outer fringes of the Lake District, and from the early 1880s the Wordsworth Society was to provide an enduring core around which opponents of further intrusive development could organise, whether it took the form of railways, water extraction, seaplanes, mining, quarrying and, eventually, speedboats and road improvements. When campaigns against such development came hard on each other's heels from the mid-1870s onwards, the Lake District Defence Society (founded 1883) and successor organisations marshalled largely external opposition (abetted by the high-class tourist trade) from academics, professionals and industrialists who wanted to keep the area quiet, 'unspoilt' and in tune with contemplative walking, and whose arguments laid emphasis on the links between literature, landscape, an imagined virtuous rural society and 'Nature and Nature's God' in constructing a defensive rhetoric which also set democratic store by the preservation of footpaths and common land.[49] The campaign of 1876 against the extension of the Windermere branch railway to Ambleside and beyond claimed the direct support of John Ruskin, by this time firmly ensconced at Brantwood on the eastern shore of Coniston Water, and incubating a reputation as combative social moralist alongside his more comfortable-seeming role as critic of art and architecture.[50] He was soon to withdraw from active public campaigning in the Lakes, but the torch was taken up by his disciple Canon Hardwicke Rawnsley, who became the Rector of Crosthwaite, near Keswick, and in 1894–5 was one of the founders of the National Trust. This organisation was to become an immensely powerful influence on the Lake District in the twentieth century, especially through the acquisition of land by purchase or gift to reserve its special qualities from development, in the furtherance of civilised common enjoyment. This theme will be developed, and placed in an international context, in Melanie Hall's chapter.[51]

The prevailing (but contested) perception of the Lake District as a place with special qualities, and in need of special protection, endowed it with a distinctive

---

47   See above, note 28.

48   Wordsworth, *Guide*.

49   Marshall and Walton, *Lake Counties*, chapter 9.

50   Hanley and Walton, *Constructing Cultural Tourism*, pp. 135–7, 191–4.

51   J.K. Walton, 'Canon Rawnsley and the English Lake District', *Armitt Library Journal* 1 (1998), pp. 1–17; and see above, chapter 1.

moral geography in ways that were both inclusive and exclusive.[52] Since the late eighteenth century it has been particularly identified with an aesthetic that links walking, contemplation, landscape appreciation (preferably informed by direct physical experience of its undulations), and literary production, originally among a self-conscious intellectual elite, but spreading among a wider constituency through and beyond the nineteenth century.[53] There was, in turn, nothing 'natural' about this: as Wordsworth recognised, it required gradual acculturation to landscape and walking practice, and it was identified with particular locations, among which the Lake District was (as Phil Macnaghten and John Urry suggest) the first 'natural' environment to be 'tamed for aesthetic consumption', 'nature turned into spectacle'.[54] The transition from contemplative tramping to the gregarious sustained physicality of hiking, which emerged in the late nineteenth century and accompanied the rise of the 'outdoor movement' during the inter-war years,[55] was followed by a growing concern to discipline and invigilate the behaviour of walkers, which the advent of the National Park (with its wardens) and the 'Country Code' intensified. The concentration of walking routes into a limited array of prescribed or preferred paths, and the commercial development and marketing in the late twentieth century of specialised clothing and equipment for recreational walkers, presented challenges to the purity of the original ideals.[56]

The continuing popularity of the illustrated walking guides by Alfred Wainwright raises interesting issues here. Wainwright, who became the Borough Treasurer of Kendal, began publishing his detailed illustrated manuscript volumes in 1955, moving steadily from local to national visibility and becoming a television personality in the mid-1980s. The books became treasured possessions. They provide a classificatory grid and a vehicle for collecting routes and comparing experiences, which goes beyond the influence of the less detailed and prescriptive information provided by earlier guide-books on the Victorian model. The seven areas into which they divide the Lake District have influenced subsequent perceptions. A Wainwright Society was founded in 2002, and significant numbers of people have begun to 'collect' the routes and summits he presents. On the one hand, this promotes appreciation of the fells and gives a sense of drive and purpose

52   D. Matless, 'Moral geographies of English landscape', *Landscape Research* 22 (1997), pp. 141–55; C. Brace, 'A pleasure ground for the noisy herds? Incompatible encounters in the Cotswolds, 1900-1950', *Rural History* 11 (2000), pp. 75–94; B. Anderson, 'A liberal countryside? The Manchester Ramblers' Federation and the "social adjustment" of urban citizens, 1929–1936', *Urban History* 38 (2011), pp. 84–102.

53   A.D. Wallace, *Walking, Literature and English Culture* (Oxford: Clarendon Press, 1993); R. Jarvis, *Romantic Writing and Pedestrian Travel* (London: Macmillan, 1997).

54   J. Urry, *Consuming Places* (London: Sage, 1995), chapter 13; P. Macnaghten and J. Urry, *Contested Natures* (London: Sage 1998), pp. 114, 123, 203.

55   H. Taylor, *A Claim on the Countryside* (Edinburgh: Keele University Press, 1997).

56   Macnaghten and Urry, *Contested Natures*, p. 203; Anderson, 'A liberal countryside'; M.C. Parsons and M.B. Rose, *Invisible on Everest* (London: Northern Liberties Press, 2003).

to those who need this kind of guidance; on the other, it encourages an obsessive frame of mind which is more concerned with the competitive accumulation of routes than with aesthetic appreciation and contemplation, channels exploration into designated paths in ways that are reminiscent of the 'stations' of the Picturesque, and is likely to exacerbate existing problems of footpath erosion. Wainwright's friend Harry Griffin, a less 'driven' enthusiast who wrote evocatively about the high fells, pointed this out to him.[57]

There is, moreover, another side to the construction of place identity around the contemplation of landscape in tranquillity, as Welberry pointed out in proposing her alternative genealogy of the idea and realisation of the National Park.[58] The Lake District has also been a venue for challenging, indeed 'extreme', sporting activity, which does not always coexist comfortably with the picturesque, the romantic or the sublime. Macnaghten and Urry have usefully drawn attention to the range of popular attitudes that come into play on these issues, and the ways in which they vary according to whether rhetorical priority is accorded to quiet recreation and the contemplation of beauty, to 'escape and freedom', or to economic development.[59] The association between the Lake District and climbing, which has generated particularly strong connections between Manchester and Wasdale since the later nineteenth century, but has more generally developed attachments to particular localities in Lakeland among members of the industrial middle (and later working) classes through both the 'conquest' and appreciation of crags and mountains, is examined in Jonathan Westaway's chapter. The dominant concern of most climbers to avoid environmental damage should be emphasised here, together with the inherent compatibility between climbing and other ways of enjoying Lakeland in tranquillity. Nor does cycling of any kind pose a direct threat to contemplative pleasures, apart from the recently recognised need to control the tendency of mountain-bikes to exacerbate the erosive tendencies of large numbers of pairs of boots on Lakeland paths. The internal combustion engine, especially in off-road incarnations, has been less readily assimilated, generating confrontations over what forms of traffic should be permitted on 'green roads'; but perhaps the bitterest controversies have involved the attractiveness of the larger lakes to power-boat and water-skiing enthusiasts. There were already fierce debates over seaplanes at Windermere in 1912, and during the inter-war years motor-boats and water-skiing took root on the largest lakes, with world water speed record attempts on Coniston Water and Ullswater as well as Windermere. Developments accelerated over the post-war generation, and in 1970 the powerboat fraternity set up an annual Records Week competition on Windermere. The Lake District Special Planning Board, and its successor the National Park Authority, made their opposition to power-boating increasingly clear (it was 'at odds with ... [the

---

57   The best introduction to Wainwright is still H. Davies, *Wainwright: the Biography* (London: Michael Joseph, 1995).

58   See above, note 20.

59   Macnaghten and Urry, *Contested Natures*, p. 197.

Authority's] ideal of providing enjoyment for walkers, yachtsmen and anglers', and a physical threat to swimmers, yachts and rowing boats), and a long Public Enquiry in the mid-1990s was followed (at some distance) by the imposition in 2000 of a prohibitive 10 miles per hour speed limit, which was eventually enforced in 2005.[60] These and related issues will be taken further in Mike Huggins' chapter.

This continuing struggle illustrates the intensifying conflicts between attitudes to the enjoyment of the Lake District which may coexist in many individuals, but are mutually exclusive where peaceful contemplation and walking in tranquillity are the defining experiences. The politician, novelist and diarist Chris Mullin expressed the strong feelings evoked by this issue when, as the junior minister responsible, he signed off the introduction of the Windermere speed limit in January 2001:

> A tiny victory for civilisation over barbarism. Unbelievably it is seven and a half years since the matter was first referred to the Department. Even now the by-law which I have ratified won't come into force for another five years. The power boaters are such a mighty vested interest that they have everyone running scared. ... But one day – five years hence – I shall be able to look out over the tranquillity of Windermere, *sans* power boats, and say 'I did that'.

What he did, in fact, was to show the necessary strength in office to put the finishing touch to years of work by campaigners; and the Manichean language of this published diary extract reflects the persisting power of association between calm, contemplation and Lakeland landscape. The predicted meltdown of a local economy without the roar of speeding powerboats did not come to pass, and training in water-skiing was able to continue on Windermere.[61] It is interesting in this context that Arthur Ransome, invoked by Karen Welberry as representing an alternative evaluation of the Lakes as site of adventure and self-realisation, attacked speeding motor-boats on Windermere in the 1930s in his *Manchester Guardian* fishing column, as Welberry herself points out. His vision of the Lakes might recognise alternative pasts and propose alternative imaginaries, but his preferred adventures were quiet, low-key and did not threaten the ideal of Wordsworthian tranquillity, with which they could coexist. Power-boats and fishing, on the other hand, made uncomfortable neighbours.[62]

The Lake District is clearly not a 'natural' landscape in the sense of a wilderness untouched by human hand, which would in any case be an impossibly idealist construct, certainly in Britain. Here the survival of such a virgin landscape, if it could be found, would necessarily be predicated on political decisions about the

---

60   O'Neill, 'Visions of Lakeland'; Walton, 'Windermere tourist trade', pp. 36–7; *Independent* (1 March 2000); http://www.conistonpowerboatrecords.co.uk/event_history. shtml (accessed 26 April 2011).

61   C. Mullin, *A View from the Foothills* (London: Profile Books, 2010 edn), p. 68.

62   Welberry, 'Arthur Ransome', p. 93.

enforcement of continuing exclusion. Such intervention would itself amount to a kind of artificial construction. If the Lake District is to lay claim to UNESCO inscription, it must be under the recently constructed category of 'cultural landscape', which was in fact created with this special case in mind. Susan Denyer discusses these important issues in Chapter 1, above. This relatively flexible concept recognises the need to accommodate changes that are in keeping with the existing character of place and people, accepts complexity, and does not require exact conformity to an externally prescribed vision. But it is usually assumed that the core of such a designation would have to lie in the distinctiveness of the post-Wordsworthian vision of the Lake District as sacralised literary landscape, as a location set apart to provide opportunities for contemplation and meditation in picturesque and sublime settings which conjure up idealised imaginings of simple, virtuous society. This need not exclude recreation, adventure and conventional forms of tourism, which are all themselves part of those aspects of the Lake District which have evolved in step with the growth of tourism (alongside, and interacting with, other industries) over the last quarter of a millennium. It is the Lake District of the National Trust, which has become such a huge landowner in the region and has the difficult task of balancing access (and car park provision) with conservation (which itself entails sustained intervention), that forms the core of any claim to 'Outstanding Universal Value'. This is also the Lake District of Wordsworth's (and, for example, Ruskin's) self-conscious torch-bearers in the various conservation, preservation and amenity societies, from the Lake District Defence Society to the Friends of the Lake District and onwards. It is, of course, since 1951 the Lake District of the National Park, whose dominant values reinforce this version of the Lake District as a 'sort of national property' by focusing on what is seen to be special about it, while finding room for alternative visions with a measure of justificatory popular appeal. Significantly, the proposed World Heritage Site boundary is also that of the National Park, as in the case of all such designations the product of sustained negotiation and compromise between the various interested parties.[63] But it is also the Lake District of (for example) the Keswick Convention, annually pursuing 'practical holiness' since 1875 in what are seen as these conducive surroundings.[64]

This version of the Lake District has to negotiate with those who want to make a living within the National Park and the potential World Heritage Site, whether as farmers, retailers, accommodation providers or managers of tourist attractions, and who frequently express their resentment at restrictions on their ability to pursue and grow their businesses.[65] But an alternative, or complementary, justification of

---

63   J.K. Walton, 'National Parks and rural identities: the North York Moors', in M. Tebbutt (ed.), *Rural Life in the Nineteenth and Twentieth Centuries: Regional Perspectives* (Manchester: Conference of Regional and Local Historians, 2004), pp. 115–31.

64   J.C. Pollock, *The Keswick Story* (London: Hodder and Stoughton, 1964).

65   P. Renouf, 'People in the Park: how National Park status threatens Lakeland communities', www.amblesideonline.co.uk/people.html (accessed 1 June 2011).

'Outstanding Universal Value' has recently put the survival of the Lake District's hill farming system at the core of its argument. Terry McCormick and colleagues argue that underlying Wordsworth's landscapes, and of more genuine universality in their relevance and message, are the practices of commons management that are essential to their existence and to an outstandingly distinctive persistent way of life (as they point out, it does not have to be unique). Successfully managed commons may nurture 'republics of shepherds', but their practical and historically informed rebuttal of the damaging assumption that competitive individualism is inherent in human nature might be thought more universally important than the high culture of Romanticism and the Picturesque, with its connotations (for some) of old-fashioned elitism.[66] This reassessment brings the Herdwick sheep beloved of Beatrix Potter, and 'hefted' or 'heafed' to their own hillsides, to centre stage in the World Heritage bid.[67] Angus Winchester's chapter is highly relevant to this theme. But the central paradox remains the Lake District's dependence on a tourism which is inherent in its modern nature, but also threatens to 'spoil' it in the eyes of people who are themselves tourists. This paradox goes back to Wordsworth himself. Such tensions would continue within a designated World Heritage Site, which would have to provide appropriate flexibility for negotiation without compromising the core identity which will inevitably be referred to as the global brand. Such conflicts and compromises are inherent in the existence of the National Park, and would remain endemic whether or not the Lake District became a World Heritage Site, as the chapters that follow will indicate very clearly.

---

66   T. McCormick, 'Wordsworth, hill farming and the making of a cultural landscape', www.cumbrianhillfarming.org.uk (accessed 1 June 2011); 'The Lake District pastoral farming system's contribution to the World Heritage Site submission', http://www. rebanksconsultinglimited.com/resources/Farming%20Contribution%20to%20WHS%20 Case.pdf (accessed 1 June 2011).

67   G. Brown, *Herdwicks: Herdwick Sheep and the Lake District* (Kirkby Stephen: Hayloft, 2009).

Chapter 3

The Landscape Encountered by the First Tourists

Angus J.L. Winchester

a white village with the parish-church rising in the midst of it, hanging enclosures, corn-fields, & meadows green as an emerald, with their trees & hedges & cattle fill up the whole space from the edge of the water & just opposite to you is a large farm-house at the bottom of a steep smooth lawn embosom'd in old woods, which climb half way up the mountain's side, & discover above them a broken line of crags that crown the scene.[1]

The landscape of Grasmere that so enchanted the poet Thomas Gray during his visit to the Lake District in October 1769 was peopled and pastoral. Like other early visitors, Gray was awe-struck by the 'turbulent chaos of mountain behind mountain' (as he described the scene in Borrowdale), but it was the juxtaposition of a domesticated landscape of pastoral farming and the wild grandeur of the fells that made Grasmere to him a 'little unsuspected paradise'. This chapter aims to sketch out the processes by which the eighteenth-century Lakeland scene had been created, for, even when Gray stumbled on Grasmere, this was an ancient cultural landscape, the culmination of many centuries of human life and labour.

When the first human eye watched spring sunlight scudding along a craggy fellside or caught a glimpse of the crystal reflections on a lake's surface on an autumn morning can only be surmised, but the tangible legacies of Neolithic peoples (the stone axe factories on the Langdale Pikes; the stone circles at Castlerigg and Swinside, for example) demonstrate that this has been a lived-in landscape for more than 5,000 years. Successive generations have left their mark, directly or indirectly, since prehistoric times, giving the landscape a time-depth stretching far into the past. Evocative though the tangible monuments created by prehistoric peoples are, the greater legacy from that period in the landscape today lies in the impact their activities had on soils and vegetation. The Bronze Age (c.2000–c.500 B.C.) appears to have been a key period in creating the landscape of the fells: the cairnfields and settlement sites which scatter the western margins of the fells in places such as Barnscar, near Muncaster, Low Longrigg on Eskdale Common, and Stockdale

---

1   *Thomas Gray's Journal of his Visit to the Lake District in October 1769*, ed. W. Roberts (Liverpool: Liverpool University Press, 2001), p. 88.

Moor and Town Bank[2] seem to be associated with pollen evidence for a significant reduction in tree cover, creating an open landscape which became vulnerable during the wetter climatic conditions that arrived in the middle of the first millennium B.C.[3] The result was to generate the waterlogged, podsolised soils on the peaty plateaux of the lower fells, creating an environment which could no longer sustain intensive use. Further clearance and settlement took place in the valleys during the Romano-British period. However, it is not at all clear how much of the later cultural landscape of the Lake District – its farms and fields, tracks and walls – is inherited from this prehistoric activity. Deserted 'native' settlement sites, probably dating from the Iron Age or Romano-British period, such as those at Lanthwaite Green, on the flanks of Grasmoor, on Aughertree Fell, near Uldale, and at Borrans near Staveley,[4] bear little relation to later patterns of settlement and land use. Indeed, the constituent elements of the landscape encountered by the first tourists – the farms, fields, tracks and woods – many of which survive today, were largely the creation of the last millennium, so the focus of this chapter is on the medieval and early modern centuries.

Colonising the Lakeland Dales

For much of the past century, a dominant assumption in the history of the Lake District has been that the Scandinavian colonists, who arrived in the tenth century, played a key role in the spread of settlement up the Lakeland valleys. The wealth of Old Norse elements in Lake District place-names and the rich Scandinavian legacy in Cumbrian dialect were the key evidence for this view, which gained wide currency through the writing of W.G. Collingwood and appeared to be strengthened in the 1960s by the results of pollen analysis which showed a phase of woodland clearance in the earlier medieval centuries.[5] Collingwood's graphic description of Viking colonisation became hugely influential. Drawing parallels with the recorded colonisation of Iceland, he wrote:

> the colonists began by settling outside the district and gradually worked up into the fells. ... a chief established himself in a 'baer' or 'by' in a readily accessible

2   Barnscar: Lake District National Park Historic Environment Record [LDNPHER], no. 4718; Stockdale Moor, Low Longrigg and Town Bank: National Trust HBSMR, nos 21943, 24370, 24397 (both available through http://ads.ahds.ac.uk/catalogue/).

3   W. Pennington, 'Vegetation History in North-West England: a regional study' in D. Walker and R. West (eds), *Studies in the Vegetational History of the British Isles* (Cambridge, 1970), p. 72.

4   LDNPHER, nos 1091, 1907, 3036.

5   For Collingwood see M. Townend, *The Vikings and Victorian Lakeland: the Norse Medievalism of W.G. Collingwood and his Contemporaries*, Cumberland and Westmorland Antiq. and Archaeol. Soc. Extra Series 34 (Kendal, 2009). For the 'Viking period' clearance phase see Pennington, 'Vegetational history', p. 74.

spot: his thralls took the cattle and sheep to summer pastures further inland and these in time became fixed farms; and his descendants cut up the original great land-take with its back-blocks until all available ground was covered.[6]

Yet there are good grounds for thinking that Collingwood's image of Viking land-taking was too simplistic. First, place-name evidence suggests that parts of the later-medieval settlement pattern of the dales had already been established in pre-Viking times; second (and probably of greater significance), the documentary record suggests that many Lakeland farms and hamlets probably date not from Viking times but from the period of sustained population growth between *c.*1150 and *c.*1300. The earlier view which interpreted the prevalence of Scandinavian words in dialect and place-names as evidence of Viking colonisation failed to take account of the fact that once Scandinavian elements were embedded in the language of the people, they persisted. The place-name element 'thwaite' (Old Norse *þveit*, 'a clearing'), for example, continued to be used to create new place-names in the thirteenth century and its meaning was still understood *c.*1600, when the antiquary John Denton could write, 'wee yet call a great plaine peece without bushes a thwaite of land, if it be severed by inclosure'.[7] So, in sketching out the processes by which the patterns of settlement and land use encountered by the first tourists evolved, we need to take a wide view, from pre-Viking times to the eighteenth century.

Few examples of pre-Viking settlement names – such as those incorporating the Old English *–tun* (meaning 'farmstead, village or estate') – are found in the Lake District proper. Such names occur mostly on the margins of the hills: Bampton, Barton, Hutton, on the eastern flanks of the Lakeland dome; Murton, Kelton, Irton, on the west; Broughton, Ulverston, Colton on the south. However, some early cores of settlement within the Lakeland dome are suggested by the place-name evidence, particularly where valley floors widened out to provide a core of cultivable land, as in the vicinity of Keswick (Old English *cēsewīc*, 'dairy farm') and Lorton, or at Coniston. Scandinavian settlement names are also rare: the element *–by* (equivalent to the Old English *tun* in meaning) is almost completely absent from the Lake District itself. The names of most farms, hamlets and villages may be classed as topographical names, recording some aspect of the landscape in which they lie. Many are derived from Scandinavian elements and particular categories of names, such as names recording the clearance of woodland (of which *–thwaite* is perhaps the most characteristic) and those referring to pastoral farming, are frequent. Dating the foundation of the farms and hamlets bearing these names is difficult: many were not recorded until *c.*1200 and only rarely can the process of settlement be reconstructed from documentary sources. Where medieval rentals or surveys of the Lakeland valleys survive, they suggest

---

6   W.G. Collingwood, *Lake District History* (Kendal, 1928 edn), pp. 42–3.

7   *John Denton's History of Cumberland*, ed. A.J.L. Winchester, Surtees Society 213 (Woodbridge, 2010), p. 37.

that much of the post-medieval settlement pattern – the location and siting of farmsteads and hamlets – had come into being before *c.*1320. The critical centuries of colonisation and the establishment of the settlement pattern thus fell between the arrivals of Scandinavian settlers in the tenth century and the later thirteenth century and much is probably to be dated to the century and a half from 1150 to 1300. Several themes can be identified to help reconstruct the processes by which the Lakeland valleys were colonised and populated.

The first is the evidence that the fells and dales provided a valuable grazing resource even before permanent settlements came to be established. Collingwood's image of 'thralls' taking livestock into the valleys to graze in the summer months is almost certainly valid. Two categories of place-name enable us to catch a glimpse of the pastoral use of the Lakeland valleys before the pattern of farmsteads and fields were established. The first are Scandinavian terms recording summer pastures: the Gaelic-Norse *erg* or *aergi*, which occurs in names such as Mosser, Sizergh, Torver and Winder, and *saetr*, which is preserved in names such as Satterthwaite, Setmurthy and Ambleside. Both terms refer to a grazing ground, the former seemingly on lower land on the fringes of the fells; the latter perhaps on higher summer pastures. The second group are those containing the Scandinavian words for pigs (*svín* and *gríss*), which probably record the grazing of pigs in areas of surviving woodland before colonisation. Some occur in small valleys running into the fells, such as Swindale, near Shap, Grisedale near Patterdale, and Mungrisdale and Swineside on the skirts of Blencathra. Some of the settlements bearing these names had been recorded by 1200; it seems likely that they capture a phase of seasonal use of pastures in the Lake District valleys in the period, say, 950 to 1150.[8]

Much of the transition from seasonal occupation to permanent settlement probably took place in the later twelfth and thirteenth centuries.[9] By then, almost the whole of the Lake District (in common with most other upland areas of medieval Britain) had the legal status of private hunting forest. Legacies of this survive on the modern maps in the names Skiddaw Forest, Ralfland Forest and Copeland Forest. The designations were legal rather than ecological: these were areas over which the great feudal overlords, whose estates embraced sectors of the Lakeland dome (the barons of Kendal, Greystoke and Egremont; the lords of the honour of Cockermouth and seigniory of Millom), claimed hunting rights. In contrast to the surrounding lowlands, most of the Lake District was not divided into manors held by intermediate lords but, rather, consisted of land under the direct control of the overlord. Its use, colonisation and development was thus controlled by the feudal overlords, with the result that different parts of the Lake District developed along

---

8   For the place-name evidence for settlement history, see G. Fellows-Jensen, *Scandinavian Settlement Names in the North*-West (Copenhagen: C.A. Reitzels, 1985); D. Whaley, *A Dictionary of Lake District Place-Names* (Nottingham: English Place-Name Society, 2006), esp. pp. xv–xxxi, 388–423.

9   The following paragraphs draw on A.J.L. Winchester, *Landscape and Society in Medieval Cumbria* (Edinburgh: John Donald, 1987), pp. 13–44.

different lines, depending on the policy of the lords of the estates to which they were attached. Common to most was exploitation for pastoral farming – hunting was relegated to a secondary concern – but the nature of the pastoral use varied. In some places the overlords maintained direct control by exploiting the pastures themselves; elsewhere, they granted blocks of upland pasture to monastic houses, which, again, might farm the land directly. But most often, the overlords appear to have allowed peasant colonists to carve new farms out of the valleys and to exploit the resources of woodlands, lakes and becks, generating rental income for the lords. These two aspects – direct exploitation by large landowners, whether the feudal overlord or a monastic house, and peasant colonisation – resulted in distinct trajectories of settlement history.

The scale of direct demesne stock farming in the Lake District was modest compared to that found in the Yorkshire Dales or the Bowland Fells, where whole valleys were retained in hand by a lay overlord or granted to religious houses. In the Lake District, demesne exploitation was largely restricted to the very heads of the valleys. Troutbeck Park had been enclosed and it and the dalehead above it preserved as a lordly grazing ground by 1272. In the western valleys of the Lake District, the lords of the honour of Cockermouth retained parts of the forest of Derwentfells in hand as private pastures in *c.*1270, including valleys deep in the fells (Hobcarton, Coledale and Keskadale) and the dalehead above Buttermere, where they established a demesne cattle farm (or 'vaccary') at Gatesgarth in 1280. Other dalehead vaccaries had been established by the barons of Copeland at the heads of Wasdale and Ennerdale by the early fourteenth century. Vaccaries were large single stock farms, managed by a handful of salaried staff, and exploiting the resources of the dalehead. Hay from meadows on the flat valley floor at the head of the lake kept the cattle across the winter, while they grazed the bowl of fellside pastures around the head of the valley during the summer. Retaining these areas in hand effectively excluded peasant colonists from the heads of these valleys.[10]

The role of monastic houses in developing livestock farming in the Lake District ought not to be exaggerated. Grants to religious houses were comparatively few and much more limited than in the Pennine dales, for example. Furness Abbey's holdings were by far the most extensive and included the whole of Furness Fells, in which they established stock farms ('herdwicks'). Elsewhere monastic property was comparatively limited: Furness and Fountains abbeys shared Borrowdale, in which Fountains established a vaccary at Stonethwaite; Furness acquired the wild pastures of upper Eskdale and established a vaccary at Brotherilkeld; the small

---

10　A.J.L. Winchester, 'Demesne livestock farming in the Lake District: the vaccary at Gatesgarth, Buttermere, in the later thirteenth century', *Trans Cumberland and Westmorland Antiq. and Archaeol. Soc.* 3rd ser., 3 (2003), pp. 109–18; A.J.L. Winchester, 'Vaccaries and Agistment: upland medieval forests as grazing grounds', in J. Langton and G. Jones (eds), *Forests and Chases of England and Wales c.1000–c.1500* (Oxford: St John's College Research Centre, 2010) pp. 109–24.

Cistercian house at Calder built 'dairy houses' on its hill properties in the upper Calder valley between Ennerdale and Wasdale.

Both the lay and monastic stock farms resulted in the establishment of settlements in these often remote locations, deep in the fells. Hay meadows and pastures were enclosed, probably creating the framework of the farming landscape that survives today. It is striking how large, convex, fellside enclosures are often associated with vaccary sites, as at Stonethwaite, Gatesgarth and Brotherilkeld, for example. By the late middle ages some of these demesne stock farms had been leased and sometimes subdivided to create a cluster of tenanted farms where the single lordly establishment had been. The lay overlords had often withdrawn from direct exploitation of their vaccaries by the fourteenth century. By the sixteenth, the vaccary at Gatesgarth had been subdivided into three holdings, while the four vaccaries at Wasdale head had become a community of 18 tenanted farms.

In parallel with the direct exploitation of the fells and dales by lay and monastic landlords, went a largely silent, unrecorded process of colonisation by peasant communities, who, by 1300, had carved farms from surviving woodland, cultivated the valleys, grazed cattle and sheep on the fellsides, burnt charcoal and smelted iron and built corn mills and fulling mills in the dales. In other words, by the early fourteenth century fully fledged communities inhabited most Lake District valleys. Documents occasionally allow us to recapture something of the chronology of this process, as in the following examples from the forest of Derwentfells. At Buttermere, for example, a core of farms beside the flat cultivable apron of land between the lakes of Crummock Water and Buttermere had probably been established before 1200 (there was a corn mill there by 1215), but colonisation in the later thirteenth century saw a ring of new farms being established on the lower fellsides behind the village. A similar process occurred down the valley at Lorton, where new peripheral hamlets were carved out of late-surviving woodland at Armaside and Swinside, again probably in the thirteenth century.

The final stages of colonisation appear to be charted in the small valley of Wythop, tucked away in the fells at the northern end of the forest, and in the significantly named Newlands valley. An inquiry in 1307 recorded that Wythop had been unsettled and valued only for grazing when John de Lucy had acquired it *c.*1260 but that it had since been 'built on and improved' and was worth ten times its previous value. Charters from the 1280s show John de Lucy buying out the grazing rights of neighbouring landowners as he settled tenants on his estate, and a suggestion of the valley's previous use as a seasonal summer grazing ground is preserved in the farm name Old Scales ('the old shielings' or herdsmen's summer huts) and hill name Lord's Seat ('the lord's *saetr*). A similar process appears to be recorded in the Newlands valley, where the rents paid by tenants rose steadily between 1266 and 1310 as new fields were carved from the wastes (Figure 3.1). Again, permanent colonisation seems to have replaced seasonal grazing: the name 'new lands' replaced the previous name 'Rogersett' ('Roger's *saetr*') and several of the farms bear names incorporating the element *skali* ('a hut'), probably referring to seasonal shieling huts.[11]

Figure 3.1   Newlands valley from Cat Bells
Note: A landscape of peasant colonisation, the farms and fields having been carved from the wastes of the forest of Derwentfells in the thirteenth century, as a summer grazing ground was converted to permanent settlement.
Source: © The author.

In these examples the documentary record provides a rare glimpse of what we can assume to have been a more widespread phenomenon. It is very likely that almost all the farmstead and hamlet sites running up the sides of valleys such as Eskdale, Langdale, and Longsleddale have been occupied since this active phase of colonisation, much of which probably took place in the thirteenth century. The crises of the fourteenth century (climatic deterioration, famine, cattle plague, sheep 'murrain', Scots raids – though these probably left the Lake District proper largely unscathed – and the Black Death) and the continuing economic depression in the early fifteenth may have resulted in some retreat from the margins. In 1324 substantial numbers of farms and cottages in the

---

11   The evidence from Derwentfells is discussed in more detail in Winchester, *Medieval Cumbria*, pp. 39–40, 140, 147–8.

Windermere area lay untenanted: 24 at Applethwaite (accounting for around one-fifth of the total rental income); 17 at Crosthwaite and Lyth (representing more than one-quarter of the rental income).[12] Whether these represented long-term population loss is not clear, but economic disruption in the late-medieval centuries probably caused some settlement desertion, such as the large, probably industrial, site at Scales beside Crummock Water. However, renewed pressure on land is in evidence by 1500, when new fields were again being carved from the fellsides and farms were being divided between sons to accommodate a growing population. By the sixteenth century much of the framework of the farming landscape visible today had been established. Little of the medieval fabric, either of buildings or of field boundaries, survives but the placing of farms and hamlets, roads and tracks, churches and mills across the Lake District landscape is largely a legacy of the medieval centuries.

The Farming Landscape: Farmsteads, Fields and Fells

By 1600 the documentary record allows a step-change in the level of detail that can be recaptured, both in the landscape and in the lives of the individuals who inhabited it. In most Lakeland valleys we can identify a handful of yeoman dynasties who formed a significant minority of the farming population, holding their estates across the generations – and enabling Wordsworth to claim, exaggeratedly, that 'the land, which they walked over and tilled, had for more than five hundred years been possessed by men of their name and blood'.[13] The names of some of them are preserved in the names of their farms: Middlefell Place in Langdale; Stephenson Ground, Jackson Ground and Carter Ground in the Lickle valley; Iredale Place and Jenkinson Place at the head of Loweswater, for example.[14] There is evidence for a growth in wealth in Lake District communities in the later seventeenth and early eighteenth centuries, accompanied by a measure of consolidation as smaller farms were absorbed into larger ones to create fewer, larger holdings.[15] Many of the distinctive traditional farmsteads of the Lake District took their modern form at this time. Dates and initials, carved into both the expensive 'freestone' (i.e.

---

12   W. Farrer and J.F. Curwen (eds), *Records Relating to the Barony of Kendale*, vol. 2, Cumberland and Westmorland Antiq. and Archaeol. Soc. Record Series 5 (Kendal, 1924), pp. 60, 91.

13   W. Wordsworth, *Guide to the Lakes*, with new preface by S. Gill (London: Frances Lincoln, 2004), p. 74.

14   Such farm names are discussed in A.J.L. Winchester, 'Wordsworth's "Pure Commonwealth"? Yeoman dynasties in the English Lake District, *c.*1450–1750', *Armitt Library Journal* 1 (1998), pp. 86–113, esp. pp. 105–7.

15   J.D. Marshall, 'Agrarian wealth and social structure in pre-industrial Cumbria', *Economic History Review* 2nd ser., 33 (1980), pp. 503–21; J. Healey, 'Agrarian social structure in the central Lake District, *c.*1574–1830: the fall of the "Mountain Republic"?', *Northern History* 44, 2 (2007), pp. 73–91.

sandstone) used for door and window surrounds, and elaborately carved wooden internal fixtures, especially spice cupboards and 'bread' cupboards, suggest major capital investment, if not a 'great rebuilding', in the period 1660–1740. Vernacular farmhouses – typically long, low and whitewashed, nestling into a sheltered spot and embowered by sycamores – often carry evidence of having been rebuilt at this time, even if the site of the farmstead had been occupied since the medieval period.[16]

The farmsteads were the hub of a working landscape, many elements of which survive today. The key division of the Lakeland farming landscape was between the *inbye* land in the valley-bottom land close to farmstead, much of it improved and used for crops, and the open unenclosed hillsides, used as common grazings. These were complementary elements in the farming system: the inbye land producing crops of corn and hay in the summer months, while the livestock grazed the open fells. The physical boundary between the two types of land was the *head-dyke* or *ring garth*, the permanent enclosure separating the common grazings from the farmland and thus keeping livestock out of the growing crops. In some medieval communities in uplands, this would have been almost the only physical boundary in the landscape. Where the fells remain open, surviving head-dykes are sometimes normal drystone walls but some take the form of a massive earthen bank, faced with stone, which would have been topped with a 'dry hedge' of brushwood (Figure 3.2).[17]

The farmsteads lay within, or were sometimes strung out along, the head-dyke, each holding a share of the precious farmland along the valley. Sixteenth- and seventeenth-century documents show that the layout of farms varied widely in different parts of the Lake District. At one end of the spectrum were farms consisting entirely of small irregular enclosures; at the other end, holdings made up largely of shares in open fields and meadows. Of the first type was Cuthbert Bell's holding at Scales, near Lorton, which consisted in 1578 of a croft and twelve other small arable fields and hay meadows, totalling a mere 23 customary acres (approximately 37 statute acres), intermixed with land of his neighbours. By contrast, farms over the fells in the village of Braithwaite, were made up almost entirely of strips of ploughland in a small open field on the alluvial fan in front of the village, shares in open hay meadows on wetter ground on the valley floor, grazing in shared pastures on enclosed grassy hillocks, and a share in the communal fuel supply, the peat moss in a boggy hollow in the valley floor.

16   S. Denyer, *Traditional Buildings and Life in the Lake District* (London: Victor Gollancz, 1991); R.W. Brunskill, *Traditional Buildings of Cumbria: the County of the Lakes* (London: Cassell, 2002); R. Machin, 'The Great Rebuilding – a re-assessment', *Past and Present* 77 (1977), pp. 33–56.

17   Winchester, *Medieval Cumbria*, pp. 59–60; A.J.L. Winchester, *The Harvest of the Hills: Rural Life in Northern England and the Scottish Borders 1400–1700* (Edinburgh: Edinburgh University Press, 2000), pp. 52–6.

Figure 3.2   The head-dyke: a stone-breasted bank at Cold Fell
Note: This massive bank, probably dating from medieval times, still forms the boundary
between enclosed farmland (on the right) and the unenclosed common grazings of Cold
Fell (on the left). It is faced with stone on the side fronting the common and would have
been topped with a 'dry hedge' of brushwood, to prevent animals from entering the growing
crops of corn and hay.
Source: © The author.

In general, where the valley floor was wide enough to contain a sufficiently large area
of flat, well-drained arable land, there would be a clustered village or hamlet sharing
the resources of an open field, as at Braithwaite, Grasmere, Coniston and Buttermere.
Even small hamlets might have a patch of open field on the valley floor: a survey of
Grizedale Hall in Furness Fells in 1697 records the survival of such an open field
in that small, remote valley. The Hall's lands included 'dales' (shares) of arable land
and meadow in 'Graisdale Town Common fields', providing not only crop land and
hay but also the vitally important 'eatage' and ' winter gate', the grazing rights in
autumn and winter after the hay and corn had been cut.[18] Most of these open fields
and meadows disappeared in a largely silent process of division and enclosure before
1750. We may envisage a piecemeal procedure of exchange and consolidation of
strips, enabling individuals to fence their share of the field, a process facilitated by

---

18   Lancashire Record Office, DDX 398/122.

the reduction in the number of holdings as result of farm amalgamations across the seventeenth and eighteenth centuries.

Where only a narrow tongue of farmland ran up the valley floor, each farm or hamlet usually had its own crop land and hay meadows, often taking the form of separate closes. Many of these areas had probably never had shared open fields and meadows, having been created instead by a process of piecemeal assarting, especially during the colonisation of the thirteenth century. In such areas we see a 'hand-made' landscape of irregular hedged or stone-walled closes, which had evolved slowly, as successive generations of farmers added to or replaced elements of the field pattern, perhaps subdividing a field or throwing two fields together, to create the manicured landscape of small fields along the valley floors which is such a characteristic feature of the Lake District.

Meanwhile, another characteristic element in the farming landscape of the Lake District valleys – the 'intakes' on the lower slopes of fells – was added to the core of farmland. These were enclosed pastures, making use of the deeper soils on the drift-covered lower slopes of fells. They probably replaced what had been unenclosed wood-pasture in the Middle Ages, where well-drained soils provided the potential for good growth of grass, to provide grazing for cattle and horses. (As the numbers of cattle and horses declined in the twentieth century, the deeper soils of these slopes have become infested with dense stands of bracken.) The chronology of intaking is only rarely recorded but many of these fellside enclosures seem to have originated as 'improvements' or encroachments on to the edge of the common grazings in the sixteenth and seventeenth centuries. In Eskdale (Figure 3.3), the process which generated the large intakes along the lower fells can be recaptured. There, open sections of the lower fellsides had been assigned to each farm as a cow pasture, to provide grazing for milking cattle close to the farmstead, in a manor court order of 1587. The result appears to have been the 'virtual' enclosure of these hillsides, as individual farmers gained exclusive rights to sections of the lower fells. By 1701 many of the intakes in Eskdale were in existence, as a result of farmers throwing a wall around a section of the fellside which they had, in effect, appropriated as their own. Though most intakes appear to be the result of individual initiative, some were created by communal activity. One of the shared pastures at Braithwaite had been taken in by the men of the village in the 1480s; Bleak Rigg, a bank of hillside deep in the hills behind Buttermere, enclosed c.1568, was shared by nine tenants in the village in 1578. In this case, enclosure proved to be comparatively short-lived – the walls enclosing Bleak Rigg had fallen into decay by the eighteenth century and the land reverted to common.[19]

Such intakes allowed for greater flexibility in the farming system. They provided secure and, often, better-quality grazing, enabling farmers to keep

---

19   Winchester, *Medieval Cumbria*, pp. 52–4. For Eskdale, see C.P. Rodgers et al, *Contested Common Land: Environmental Governance Past and Present* (London: Earthscan, 2011), p. 96.

Figure 3.3   Intakes on Great Barrow, Eskdale
Note: Enclosed pastures on the lower fellsides, which were 'tacked on' to earlier farmland in the valley floor. The intake on the right of this photograph had been enclosed before 1578.
Source: © The author.

livestock out of the arable fields and meadows without having to turn them on to the higher fells. Rams could be kept from the ewes in autumn to control the timing of lambing; cows could be kept close to the farm for ease of milking. More flexible stock management probably enabled greater numbers of animals to be kept, in turn leading to an increase in the wealth of Lake District farming communities between *c.*1650 and *c.*1750.[20]

The open fells beyond the intake walls also formed a vital part of the farming landscape, complementing the farms and fields in the dales.[21] First and foremost, the fells provided thousands of acres of grazing ground for livestock in the summer months. Though much of it was poor ground, craggy or waterlogged, the thin pastures enabled farmers to keep the flocks and herds on which their livelihood was founded. The fells also yielded a range of other essential resources: peat, the principal fuel

---

20   Marshall, 'Agrarian wealth and social structure'; Winchester, *Harvest of Hills*, pp. 68–73, 146–8.

21   The following paragraphs draw on Winchester, *Harvest of Hills*, passim. For a case study of the use and management of the fells, see the discussion of Eskdale Common in Rodgers et al, *Contested Common Land*, pp. 89–99.

in most Lake District valleys until the nineteenth century; bracken, the traditional thatching material and source of litter for livestock bedding in uplands where there was little straw; rushes, for strewing on floors and as the wicks for rushlights. This variety of resources was exploited through the exercise of common rights.

The open fells were legally 'manorial waste', land owned by the lord of the manor but over which his tenants held common rights allowing them to use the resources of the common. The legal status of parts of the fells as the lords' private pastures – a legacy of their status as medieval hunting forests – survived in some places, such as Stockdale Moor, between Ennerdale and Wasdale, but the fells surrounding most valleys were treated as common grazings by the sixteenth century. Common rights were attached to holdings of land in the dales, so that a farmer could exploit the resources of the fells only to support his farm: he could put on to the fells in summer only the livestock he could keep over winter on his farm and could gather only so much bracken or peat as was necessary to use on his farmstead. As with all communal resources, tensions could arise between individual interests and the common good and these were negotiated and resolved by the manor court, a 'local parliament' which normally met twice a year, in spring and autumn, under the auspices of the lord of the manor's steward. All tenants of the manor were expected to attend and decisions were taken by a jury drawn from the tenant community. Among the court's various functions was resolving disputes and formulating and policing byelaws (or 'pains') to govern the exercise of common rights. In seeking to manage the demands on the resources of the fells and to maintain good neighbourly relations, the courts often devised management rules of some sophistication. Sections of the common might be allocated to individuals: a farm would have its own 'peat pot' in which to dig fuel, its own 'bracken room' or 'dalt' from which to gather thatch and litter, and its own sheep heaf, on which to graze its flock. Explicit limits on the quantity of a resource which could be taken might be set (so many 'dayworks' of peat or bracken, for example) and seasonal restrictions were sometimes imposed, particularly where a resource was subject to conflicting demands. Bracken, the scourge of the modern hill farmer, was a case in point. Not only was it used for thatching (before slate roofs became common from the seventeenth century) and as litter for animal bedding, it also gained a value as a cash crop when burnt into ash. Sales of bracken ash as a source of potash for soap-making and the glass industry were an important source of revenue for some Cumbrian farmers in the seventeenth and eighteenth centuries and – though burning bracken for sale broke the fundamental principle that common rights could be exercised only to support the holding to which they were attached – the manor courts accepted this activity and devised rules to accommodate it. They often fixed a 'bracken day', usually around Michaelmas (29 September), before which bracken was not to be mown for litter or for burning, and some courts set another date a few weeks earlier, after which bracken could be cut for thatching, thus allowing the careful cutting or reaping of fronds needed for one purpose to be carried out before the wholesale clearance of the plant later in the season.

The open fells were thus a hive of activity, their use and exploitation generating a seasonal cycle of coming and going. Sheep, wintered on enclosed fields or on the common close to the farm, were driven up to their heaf in the spring, the ewes and lambs following the wethers (castrated males) once lambing time was over. Meanwhile, in early May, peat would be dug from the 'peat pots' and left to dry. Midsummer saw the flock gathered in for washing and clipping as the valuable annual crop of wool was taken, then driven back to the heaf. In late July or early August, cattle grazing on the lower fells would be brought back into the inbye land to graze on the 'fog' or aftermath of the hay meadows. The dried peats would be either carried down to the farm (on pack pony or by sledge) or taken to a 'peat scale', a small drystone hut on the fellside, where they would be stored until needed. In September, as we have seen, bracken would be harvested and at the end of the month smoking piles of the plant would dot the common.[22] In November the sheep were again gathered in, for 'salving', a tedious task when a mixture of butter and tar was spread on the skin to keep down parasites, and sorting. The 'hoggs' (the lambs of the previous spring) would then be put to better grazing for the winter, either in an intake with a hogghouse for shelter, in which hay was stored, or by being sent to winter away in the lowlands surrounding the Lake District. The rams ('tups') were then put to the ewes to restart the cycle.

The repeated to-ing and fro-ing from farmstead to fell meant that the fellsides were a peopled landscape. Sixteenth- and seventeenth-century documents record the part played by youths in herding and tending livestock on the commons, while the back-breaking toil required to dig and stack peat and to harvest bracken and rushes would have involved all generations of a farming family. The fells still carry physical evidence of these patterns of traditional use: sheepfolds in which to gather and sort the flock; washfolds, placed beside a deep pool in a beck in which the sheep were washed before clipping (Figure 3.4); worked out peat diggings, often now no more than a waterlogged depression or a small body of standing water; clusters of peat huts, such as those on Boot Bank in Eskdale or beside Rowantreethwaite Beck on Mardale Common, often beside well-made packhorse or sledge tracks leading down the steep fellside to the farmstead. All these are reminders that the open fells were an integral part of the farming landscape, complementing the fields in the valley bottom.

---

22   There is very little evidence that bracken was burnt in kilns: the supposed 'bracken-burning kilns' identified by the late Michael Davies-Shiel – M. Davies-Shiel, 'A little-known late mediaeval industry, part I. The making of potash for soap in Lakeland', *Trans. Cumberland and Westmorland Antiq. and Archaeol. Soc.* new ser., 72 (1972), pp. 85–111 – are more likely to have been kilns in which wood was burnt, also for potash.

Figure 3.4   Washfold beside Worm Gill, Stockdale Moor

Note: A small fold with a narrow exit into a deep 'wash dub' in which sheep would be washed before shearing. This is a humble example (possibly of medieval date) of the washfolds which continued in use on most Lakeland farms until the nineteenth century.
Source: © The author.

Industry and Woodland

The landscape of pastoral farming was but one element of the scene encountered by the early visitors. The Lake District had also long been a landscape of industry and, while the scale of industrial operations would explode in the nineteenth century, most valleys would have borne evidence of industrial activity in the eighteenth century. Mining and quarrying to capture the mineral wealth of the fells had a long history, though the scale of later exploitation has often dwarfed, if not obliterated, evidence of early activity. Non-ferrous ores had been mined since the medieval period (lead mining is recorded in Derwentfells from the thirteenth century) and underwent a significant expansion in the later sixteenth century when the Society of the Mines Royal brought in miners from Germany to win lead, copper and other minerals from the fells around Caldbeck, Newlands and Coniston.[23] Roofing slate had also been won from the hills since at least the seventeenth century: Skiddaw slates were replacing thatch as the roofing material for farmhouses by the later

---

23   C.M.L. Bouch and G.P. Jones, *The Lake Counties 1500–1830: a Social and Economic History* (Manchester: Manchester University Press, 1961), pp. 119–27.

seventeenth century and even modest Lakeland farmhouses were slated by the time the first tourists arrived in the 1760s.[24]

Textile manufacture also had a long history in the area. Cumbria was one of the first regions of England to adopt water-powered fulling mills to 'walk' and thicken woollen cloth. They are frequently recorded in surveys from the century 1250–1350, confirming the presence of a local cloth-weaving industry in medieval Lakeland. In the south of the region, cloth production and finishing had become concentrated in the towns by 1500, 'Kendal green' being famous as the coarse cloth of the labouring man. Spinning and some weaving continued to be carried out in the countryside (witness the open 'spinning galleries' in some Lakeland farmsteads) and local specialities could be found, such as 'Skiddaw grey' ('a good wearing cloth') the production of which was focused on fulling mills in the Bassenthwaite area in the 1680s.[25]

One important element in the landscape of the Lake District valleys was intimately linked to industrial activity: the broadleaved woodlands in walled enclosures on the lower fellsides. As elsewhere in Britain, these are ancient features of the landscape, rooted in the medieval past. They formed an important resource in the traditional farming economy but their distribution and survival owed much to the part they played in early-modern industry, reminding us that the Lakeland landscape was also a product of economic forces other than livestock farming. The distribution of semi-natural woodland in the Lake District today exhibits a strong north-south contrast, most woodland surviving in the southern parts and reflecting the concentration of woodland-demanding industries in southern Cumbria, particularly the Furness area, from the late-medieval centuries. The key to the distribution lies in the iron industry, which saw the establishment of charcoal-fuelled forges and blast furnaces across southern Cumbria in the seventeenth and eighteenth centuries. Coppice woodland supplying charcoal for the iron industry was thus a key feature of the economy of this part of Cumbria and, even as the number of ironworks declined in the nineteenth century, demand for poles remained buoyant in the bobbin mills turning bobbins for the Lancashire cotton industry.

Taking the long view of the history of Lake District woodlands, we can conceive of an evolving contest between the place of woods in the farming economy and demand for wood for industrial purposes. Access to woodland was controlled by the customary law of the manor in the late-medieval and early modern centuries. In a legacy of the status of most of Lakeland as medieval hunting forest, woodland was deemed to belong to the lord of manor, even when growing on a tenant's farm. On many manors a distinction was drawn between '*woods of warrant*', the more valuable timber, to which tenants had strictly-controlled and limited access, and

24 Thomas Denton recorded houses 'covered with Skydew blew slate' in 1687: Denton, *Perambulation*, p. 146. Many farmhouses in the western Lake District were described as 'slated' in 1758: Cumbria Record Office, D/Lec, box 300, Brown's survey, 1758.

25 Winchester, *Medieval Cumbria*, pp. 117–20; Bouch and Jones, *Lake Counties*, pp. 132–41; for Skiddaw grey, see Denton, *Perambulation*, p. 140.

underwood to which they had more general rights. The woods of warrant (defined in Kendal barony as oak, ash, holly and crab apple) were controlled by the lord's bailiff, to whom tenants had to apply for timber for the repair of their houses, for example. On most manors, tenants could take underwood (species such as hazel, alder and birch) for fuel and other necessary uses (such as obtaining 'spelks', slender wands for securing thatch), often paying a notional sum of rent for this right called 'greenhew'.[26]

Figure 3.5   Ash pollards by Newlands church
Note: Pollarded ash trees are a feature in many Lake District valleys. Pollarding allowed a crop of useful poles and nutritious leaves to be grown on trees in farmland, out of reach of grazing livestock.
Source: © The author.

But the reality was more complex and demand for wood in the farming economy sometimes brought local communities into conflict with their lords. One of the most important roles of woodland in the farming economy was as winter fodder for livestock, for which the 'croppings' of ash and holly were particularly important. It was demand for ash branches that lay behind the pollarding of hedgerow ash trees, making ash pollards one of the characteristic elements in the Lake District

---

26   Winchester, *Medieval Cumbria*, pp. 102–3; M. Parsons, 'The woodland of Troutbeck and its exploitation to 1800', *Trans. Cumberland and Westmorland Antiq. and Archaeol. Soc.* new ser., 97 (1997), pp. 79–100.

landscape (Figure 3.5). By the early eighteenth century, pollarding was viewed as being detrimental to the lord's interest. In 1718 the lord's steward reported that trees growing on the tenants' farms in the Eskdale area were 'very much abused by cutting of[f] the topes, their being not one in sixty with the top on. They are generly topt about 7 or 8 foot high. When the top is cut of[f], the tree generly takes water, and when they are watered it hinders them to thrive'.[27] Where trees growing on tenants' farms were deemed to be the property of the lord of the manor, traditional woodland management damaged the lord's rights.

Indeed, trees growing on farmland became a focus for continuing battles over custom between lords and tenants in the eighteenth century. A series of disputes in the 1750s saw escalating tension as lords harassed their tenants by exercising their rights to trees. John Pennington, the lord of Muncaster, attempted to deny his tenants' rights to 'woods of warrant', culminating in a sale of timber to two ship's carpenters from Whitehaven in 1752, which allowed them to cut down all wood growing on the tenants' farms. When it came to court, however, the tenants' rights were upheld. Further disputes arose in the Keswick area when the lord of the manor of Braithwaite and Coledale sold trees in the tenants' hedgerows for pit props in west Cumberland coalmines in 1757 and when the lord of Bassenthwaite sold all the timber on his tenants' estates for charcoal two years later.[28]

Woodland also provided income and employment to local communities quite separate from its part in the farming economy. In at least one instance woodland provided a regular market crop: nuts. Writing in 1687, Thomas Denton commented on the 'great plenty of Nuts' on the 'large hasyl trees' in Borrowdale,

> wherof the inhabitants make a profit, having a custome never to gather any nuts there under a penalty untill a certain day after Michaelmas, when they all come generally with long clubbs & beat them down from the trees (being full ripe), and then they all gather them from off the ground & sell great quantities of them in the marketts.[29]

Elsewhere in the Lake District, woodland provided the raw materials for a wide range of local by-employments. The oak woods at Thornthwaite and above Keskadale in the Newlands valley yielded bark for tanning, a local specialism which had its roots deep in the medieval centuries. In Furness Fells, probably the most wooded part of the Lake District by the end of the middle ages, a spectrum of woodland industries developed. At the Dissolution, Furness Abbey received income from craftsmen involved in 'Bastyng, Bleckyng, byndyng, making of Sadeltrees, Cartwheles, Cuppes, Disshes and many other thinges wrought by Cowpers and Turners'. 'Bastyng' was the manufacture of coarse matting and

---

27   Cumbria Record Office, D/Lec, box 169, H. Westray to W. Coles, 3 March 1717/18.
28   C.E. Searle, 'Custom, class conflict and agrarian capitalism: the Cumbrian customary economy in the eighteenth century', *Past and Present* 110 (1986), pp. 106–33.
29   Denton, *Perambulation*, p. 135.

Figure 3.6   Charcoal pitstead in Great Bank Coppice, Miterdale
Note: Circular platforms such as this, on which a stack of coppice poles would be burnt into charcoal, survive in many areas of ancient woodland. In this case, the wood has been replanted and a sturdy conifer rises from the pitstead.
Source: © The author.

swill baskets; 'bleckyng' (i.e. 'bleaching') referred to making potash for use in the textile industry; 'byndyng' was cooperage.[30]

Well before the rising demand for charcoal for the iron industry, therefore, the woods of the Lake District formed an important part of the local economy. But it was the increasing commercial demand for charcoal and the desire of landowners to increase income from their estates which led to more active woodland management and the creation of the coppice woods, walled to protect them from grazing animals, which formed an essential element of the landscape encountered by the first tourists. One of the earliest pieces of evidence for active woodland management comes from the shores of Derwentwater, where two enclosures, 'Catbelclose' and 'Scurlothyn Parke', formerly leased out as wood pastures, were taken back into lordly control to be managed for charcoal production for smelting ore from the local lead mines in 1454. By the mid-sixteenth century there is ample evidence of coppice woodland, cut on rotation, in Furness Fells and the Kendal area, and we may envisage a significant increase in such management when lead- and copper-mining expanded under the auspices of the Society of

30   A. Fell, *The Early Iron Industry of Furness and District* (Ulverston: Hume Kitchin, 1908), p. 105.

the Mines Royal from 1564. As demand for charcoal for the iron industry grew as a result of the establishment of, first, water-powered bloomery-forges in the middle decades of the seventeenth century and, second, the much more charcoal-hungry blast furnaces in the early eighteenth, so coppice management would have grown. Indeed, the acreage under trees in Furness Fells may well have grown in the eighteenth century as pastures were turned over to coppice woods. The legacy of this industrial production in the region's ancient woodlands is visible in the circular charcoal-burning platforms (or 'pitsteads') which survive in many former coppices (Figure 3.6), along with other structures such as bark-peelers' huts and potash kilns.[31]

Woodland management, whether for coppicing or timber production, involves felling trees and thus sudden and violent change to the face of the landscape. The wooded elements of the landscape extolled by the first tourists to the Lake District were therefore not static but subject to cycles of change. By the 1790s the impact of coppicing and timber extraction on picturesque scenery was being noted. Footnotes were added to descriptions of the beauty of scenery which was said to be 'cloathed in wood'. Thomas West, in the first edition of his *Guide to the Lakes* (published in 1778), had described the west bank of Derwentwater in such terms. When the fifth edition was published in 1793 a note was added: 'There is one impediment attends [West's] descriptions ... and that is the annual fall of timber and coppice wood'. By the felling of woodland on the western bank of the lake in the 1780s, wrote another commentator, 'the lake is deprived of one of its chief ornaments'.[32] This serves as a reminder that the landscape encountered by the first tourists was a dynamic one, in which change was driven essentially by the needs of the local economy. As that economy grew, so the pace of change accelerated. By the mid-eighteenth century, the Lake District landscape was on the cusp of transformation: within a few decades of its 'discovery' by the first tourists many of its fellsides were to be walled and divided in the great surge of Parliamentary enclosure; new plantations of alien conifers would be planted and the scale of mining and quarrying would be transformed. Yet, despite the changes which were about to come, the legacy of medieval colonisation and of centuries of pastoral farming continues to underpin the form and fabric of the Lake District landscape today.

---

31   Winchester, *Medieval Cumbria*, pp. 104–7; M. Bowden (ed.), *Furness Iron: the physical remains of the iron industry and related woodland industries of Furness and southern Lakeland* (Swindon: English Heritage, 2000).

32   W. Hutchinson, *History of the County of Cumberland* (Carlisle, 1794), p. 173.

Chapter 4

Landscape and Society: The Industrial Revolution and Beyond

John K. Walton

The development of industrial Britain from the eighteenth century onwards affected the Lake District both directly and indirectly; and the rise of commercial tourism was itself an integral part of the industrialisation process, as an important element of the consumer revolution which constituted the inescapable corollary of the immense expansion of the productive capacity of the economy. The technological and organisational transformations associated with the production and distribution of goods still dominate analysis and explanation, though less overwhelmingly than was the case a generation ago; and parts of the Lake District itself experienced the ebb and flow of extractive and manufacturing industries, as well as being affected by the articulation of new transport systems and the demand of industries and industrial populations for its water resources. Moreover, economic and demographic change on the Lake District's fringes also affected the views from within (especially as an industrial coastal fringe developed, stagnated and declined), and generated additional sources of local demand for the various pleasures and relaxations it offered, while improved access over extending distances widened the field of attraction to more distant urban and industrial populations. We introduced these themes briefly in Chapter 2: it is now time to develop them. But the key point is that the nature of the interaction between farming and tourism, the most traditional of productive industries and the most intangible and elusive of consumer products, was at the core of the developing identity of the Lake District as a special place from the eighteenth century onwards, even as the local agricultural systems experienced major changes which, in themselves, had little or nothing to do with tourism.

Historically the Lake District was, and remained, overwhelmingly an agricultural economy, until tourism became the leading sector in its own right in the late twentieth century. Mining, quarrying and manufacturing were important locally at times, but by the early twentieth century rural crafts were being protected and revived as part of the district's 'heritage'.[1] The nature of the Lake District's agriculture lay behind the development of its landscapes, and of the literary representations of rural society that underpinned the idea that here

---

1   J. Brunton, *The Arts and Crafts Movement in the Lake District: a Social History* (Lancaster: Centre for North-West Regional Studies, 2001).

was a special place, identified with distinctive, increasingly archaic rural virtues. The idea of the 'yeoman' or 'statesman' was central to this, and part of a wider perception that the Lake District was an 'odd corner of England' in which small independent farmers and traditional agricultural arrangements survived longer and more tenaciously than elsewhere.[2] 'Customary tenure' was an important dimension of independence, supposedly based on the holding of land in exchange for border military service against the Scots, and providing security of tenure subject to inflexible rents that were outpaced by inflation, and occasional, often more contentious payments (sometimes in kind) on the death of the landlord or transfer of the property. This severely limited the day-to-day power of absentee landowners; and substantial landed estates were anyway less in evidence here than elsewhere in England. Cumberland and Westmorland were at or near the bottom of F.M.L. Thompson's table of the 'density of country seats in 1865', and although they were not comparatively lacking in 'great estates', they were right at the bottom of the table showing the proportion of land belonging to the 'greater gentry'.[3] As we saw in Chapters 1 and 2, Wordsworth himself had propagated the ideal of the 'pure commonwealth of shepherds and agriculturalists', a self-regulating egalitarian society founded on the local democracy of the manorial court, frugal, hard-working and untainted by ambition or consumerist pretension.[4]

Historical research on the Lake District and upland Cumbria has qualified such perceptions without destroying the underlying theme of a distinctive, deeply rooted society with its own cultures and practices, grounded in hard experience of local characteristics and differences, transmitted between the generations. Indeed, this emerges clearly when we look at the results of Wynne's interviews with sheep farmers in the western uplands after the Chernobyl disaster, and the uneasy relationship between their organic local knowledge of weather, soils, vegetation, seasonality and topography, and the hard, sharply defined grids and definitions that external scientists sought to impose in ignorance or disparagement of their experience.[5] But by the late eighteenth and early nineteenth century the social landscape, and in its wake the physical one, was subject to accelerating change. The terminology of social description was shifting, but in ways that were not necessarily helpful to understanding the processes at work. A myth grew up around the alleged distinctiveness of the region's so-called 'statesmen', who were identified with the important stratum of customary tenants or otherwise independent

---

2   C. Searle, 'Custom, class conflict and agrarian capitalism: the Cumbrian customary economy in the eighteenth century', *Past and Present* 110 (1986), pp. 109–33.

3   F.M.L. Thompson, *English Landed Society in the Nineteenth Century* (London: Routledge and Kegan Paul, 1963), pp. 30, 32, 113–17.

4   T. McCormick, 'Wordsworth, hill farming and the making of a cultural landscape', *North West Upland Farming* (2009), http://www.cumbriahillfarming.org.uk/hillfarming/wordsworth.html (accessed 6 April 2011).

5   B. Wynne, 'Misunderstood misunderstandings: social identities and public uptake of science', *Public Understanding of Science* 1 (1992), pp. 281–304.

small proprietors of their family farms, whose imagined lifestyle of hard work and frugality was both pitied and admired, in an emergent genre of romantic historical topography which both mourned and celebrated gradual decline and extinction in the face of emergent modernity. Forty years ago John Marshall's demolition of the 'statesman' myth, a forensic dissection which deserves to be more widely disseminated, showed that this evocative label was almost invisible in eighteenth-century documents, that it was in any case not peculiar to the Lake District or to northern England, and that its usage in the Lakeland context had been effectively invented by the compilers of reports to the Board of Agriculture in the 1790s, followed by the topographical writer John Housman and soon afterwards by Wordsworth himself, at the beginning of the nineteenth century. The subsequent efflorescence of topographical and tourist writings on the Lake District reinforced and propagated the idea of the statesman, which was an attractive ancestry to claim; but it was a myth nevertheless.[6]

The similarly value-laden attribution 'yeoman', which was widely deployed in the eighteenth-century Lake District as across much of England, did provide a relevant form of social description, but without the same depth of celebratory romantic connotation; and it did indeed fall gradually into eclipse during the nineteenth century. If, as seems likely, the declining visibility of the 'yeomanry' in contemporary sources was due more to changing usage than to the actual demise of a social stratum, as the vocabulary of social description for family-run farms shifted from status labels like 'husbandman' and the more prestigious 'yeoman' to the modern, market-based descriptor of 'farmer', this amounted to a significant transition in perception and values, if not in actual social structure. It was aided and abetted by the conventions of census takers, who did not recognise 'yeoman' as an occupational category.[7]

The spread of competitive market values, which is what is important here, was also indicated by increasing conflict over the management of common grazing, especially as outsiders sought to exploit it; but communal values of consensual management and mutual aid remained resilient. Parliamentary enclosures eventually spread across much of the upland Lake District in the late eighteenth and early- to mid-nineteenth century, tracing straight new lines across the fell-sides and importing elements of a calculating rationality that challenged the ethos of the Picturesque or sublime, as stone walls demarcated the newly defined territories, although the crafted roughness and irregularity of their stonework gave a different impression at close quarters from the regular geometrical tracings they imposed across rough grazing, heather, gorse

---

6   J.D. Marshall, '"Statesmen" in Cumbria: the vicissitudes of an expression', *Trans. Cumberland and Westmorland Antiq. and Archaeol. Soc.* new ser., 72 (1972), pp. 249–83.

7   J.K. Walton, 'The strange decline of the Lakeland yeoman: some thoughts on sources, methods and definitions', *Trans. Cumberland and Westmorland Antiq. and Archaeol. Soc.* new ser., 86 (1986), pp. 221–33; D. Uttley, 'The decline of the Cumbrian "yeoman" revisited', idem. 3rd ser., 8 (2008), pp. 127–46.

and bracken.[8] The shared management of other resources held in common, such as peat, bracken, stone and even poor-quality coal, important for heating, cooking, animal bedding, building and roofing, was also an enduring source of tensions between communality and conflict.[9] Enclosures on the fells, where they took place, brought about a significant transformation of the upland landscape, although even after the movement had passed the Lake District was left with uniquely extensive areas of common land, especially in the most picturesque areas.

Such social structures, arrangements and relationships, and indeed rural landscapes, were not peculiar to the Lake District. They were widespread across Cumberland, Westmorland and adjoining northern counties.[10] This also applied to some of the social peculiarities which might be associated with Lakeland, without necessarily forming part of the core of the 'republic of shepherds' myth. John Marshall has commented on the interesting coincidence of comparatively low rates of recorded crime, high levels of pre-marital births to women 'out of wedlock', and high levels of basic literacy, as measured, unsatisfactorily but necessarily, by signatures and marks on marriage registers, in rural nineteenth-century Cumberland and Westmorland.[11] Similar phenomena were observable not only in upland rural northern England beyond the Lake District, wherever mining and manufacturing industry had not penetrated, but also in southern Scotland; and although they are compatible with notions of a self-policing society with strong social institutions (most of the young mothers were anticipating marriage rather than enjoying a libertine youth, and few found their way into pauperism or institutional care), the Lake District formed part of a much wider picture.[12]

---

8   Searle, 'Custom'; I. Whyte, *Transforming Fell and Valley: Landscape and Parliamentary Enclosure in North-West England* (Lancaster: Centre for North-West Regional Studies, 2005); idem., 'Parliamentary enclosure and changes in landownership in an upland environment: Westmorland, *c.*1770–1860', *Agricultural History Review* 54 (2006), pp. 240–56; idem., 'Political Spaces and Parliamentary Enclosure in an Upland Context', in B.A. Kúmin (ed.), *Political Space in Pre-Industrial Europe* (Aldershot: Ashgate, 2009), pp. 95–113.

9   A. Hillman, 'Common rights to stone, peat and coal in South Westmorland before and after enclosure', PhD thesis, Leeds Metropolitan University, 2011.

10   N. Gregson, 'Tawney revisited: custom and the emergence of capitalist class relations in North-East Cumbria', *Economic History Review* 42 (1989), pp. 18–42.

11   J.D. Marshall, 'Some aspects of the social history of nineteenth-century Cumbria (i): migration and literacy', *Trans. Cumberland and Westmorland Antiq. and Archaeol. Soc.* new ser., 69 (1969), pp. 280–307; idem., '(ii): crime, police, morals and the countryman', ibid., 70 (1970), pp. 221–46; idem., 'Out of wedlock: perceptions of a Cumbrian social problem in the Victorian context', *Northern History* 31 (1995), pp. 194–207.

12   W.B. Stephens, *Education, Literacy and Society 1830–70* (Manchester: Manchester University Press, 1989), pp. 54–7, 68; R.A. Houston, *Scottish Literacy and the Scottish Identity* (Cambridge: Cambridge University Press, 1985), p. 21.

It would also be an over-simplification to identify the 'yeomanry' exclusively with the apparently feudal survival of customary tenancies regulated by the manor court: many small to medium family farms were made up of land held on a variety of tenures, which might include parcels in outright freehold ownership, or held on fixed-term leases or rack rentals, as well as 'customary tenure'. Moreover, the manor court itself was in decline as a working institution in many areas by the late eighteenth century, and outlying farms might be several miles away from its intermittent meetings.[13] But there was also a small upper stratum of local rural society in which the substantial 'yeomanry' took on roles which the gentry might have carried out elsewhere, while developing during the eighteenth century habits and expectations about comfort, consumption and accessible luxuries which did not match the abstemious assumptions associated with a 'republic of shepherds'. The Brownes of Town End at Troutbeck, a settlement of substantial stone farmhouses close to the eastern shore of Windermere lake, exemplify this most affluent level of the yeomanry in a way that is still visible at the family home, now owned by the National Trust and open to visitors; and John Marshall and others have demonstrated that a general expansion in the range and quality of basic possessions, and in standards of comfort, was already apparent in the inventories of rural families in the southern Lake District during the first half of the eighteenth century.[14] This undoubtedly continued into and through the nineteenth century, spreading down the social scale, although anything more than modest comfort remained unsustainable even with the aid of summer 'bed and breakfast' income, which grew steadily in the railway age alongside the supply of milk to urban markets, and might be supplemented from the early twentieth century by small additional sums from campers.[15] Tourism became more important to the economies of well-placed farms as pressure on purely agricultural incomes grew, and the rural family economy remained labour-intensive, especially as out-migration reduced the supply of farm servants and labourers in an area where the system of living-in farm service (and its associated hiring fairs) remained prevalent through, and beyond, the nineteenth century. Here again, such issues were generic across much of rural northern England, and were never peculiar to the Lake District itself.[16]

---

13   Walton, 'Strange decline'; J.D. Marshall, *Old Lakeland: Some Cumbrian Social History* (Newton Abbot: David and Charles, 1971), chapter 2.

14   Marshall, *Old Lakeland,* chapters 6 and 8; L. McGhie, 'Consumer and consumption, 1650–1750: a study of household goods and the middling sort in South Westmorland and Furness', MPhil thesis, University of Central Lancashire, 2002; J.D. Marshall, 'Agrarian wealth and the social structure in pre-industrial Cumbria', *Economic History Review* 33 (1980), pp. 503–21.

15   F.W. Garnett, *Westmorland Agriculture 1800–1900* (Kendal: Titus Wilson, 1912).

16   J.D. Marshall and J.K. Walton, *The Lake Counties from 1830 to the Mid-Twentieth Century* (Manchester University Press, 1981), chapters 3–4; Marshall, *Old Lakeland*; J. Catt, *Northern Hiring Fairs* (Chorley: Countryside Publications, 1986); M. Capstick, *Patterns of Rural Development: a Study of North Westmorland and its Problems* (Kendal: Westmorland County Council, 1970).

Agriculture and rural society in Lakeland was, then, subject to increasing pressures from without and within, as industrialisation proceeded outside its imagined (pre-National Park) boundaries, and had a direct impact within them in certain areas. Before the arrival of the railways, the driving of cattle from Scotland and Ireland made its mark on Lake District geography, not least through the intermittent pressures it placed on communal grazing, as drovers (often deploying intimidating dogs) sought to refuel their hungry charges. But during the eighteenth century it also provided opportunities for Lakeland farming families to make money by providing pasture for cattle (and services for drovers) in transit, and to buy in itinerant stock to fatten on their land and sell on subsequently. Most of the main drove routes skirted the central Lake District, and the important cattle markets and fairs, such as the ones at Brough and Rosley, lay outside the future National Park; but the system, which might also be used by pack-horse trains and other commercial traffic (but remained beyond the purview of the tourist), still made its mark on topography and landscape, while helping to intensify the connections between seemingly isolated farmers, external markets, and the outside world in general.[17] It took the accumulative impact of the railways, following on from the earlier development of livestock traffic by steamer from various Cumbrian ports, to bring all this to an end by redirecting traffic flows and redefining marketing practices during the middle decades of the nineteenth century.

So, even before the literary foundations of the 'republic of shepherds' myth were being consolidated, the rural society of the Lake District was already proving increasingly permeable to outside influences, including those introduced by early tourism. The development of turnpike roads contributed to this process, partly by making the area more comfortably accessible from population centres as the roads from south Lancashire and West Yorkshire to its fringes were improved during the third quarter of the eighteenth century, although the hazardous crossing of the Morecambe Bay sands at low tide remained a popular way of reaching the Lakes until the railway opened between Lancaster and Ulverston in 1857. To a lesser extent, and later, improved roads also eased transit within the Lake District itself, although the only actual turnpike roads were those that linked Kendal with Ambleside and Keswick, Kendal with the southern edge of Windermere, and Penrith with Cockermouth via Keswick.[18] Even before the end of the eighteenth century turnpikes, tourists and trade were blamed for introducing disruptive innovations and fashions, corrupting or compromising the simplicity of rustic virtue. The story of Mary Robinson, the 'Maid of Buttermere', was a cautionary tale in several dimensions: the adolescent daughter of the Buttermere innkeeper was celebrated (not least by Wordsworth himself) as the epitome of innocent, unaffected, unspoilt rural upland beauty, only to become the victim in 1802 of a bigamous fraudster attracted by her fame, winning plaudits for her dignified demeanour in distress,

---

17   Searle, 'Custom'; Marshall, *Old Lakeland*, chapter 4.

18   L.A. Williams, *Road Transport in Cumbria in the Nineteenth Century* (London: George Allen & Unwin, 1975), pp. 10–12, 116–27.

and recovering literary celebrity through Melvyn Bragg's novel nearly two centuries later.[19] Commentators have understandably been unable to resist pointing the obvious morals and metaphors, but, as Harriet Martineau was to argue half a century later from her practical, progressive, utilitarian perspective, this loading of the dice might be misleading. She was scathing about notions of a virtuous or abstemious peasantry, preferring to highlight episodes of drunkenness, disorder and ignorance. She argued that the recent arrival of the railway at Windermere was a civilising influence on the whole area, promoting healthier living, improved education and the diffusion of basic household amenities.[20]

The heart of the Lake District was never to be penetrated by the railway system, but its tendrils reached into the area from every side. The Cockermouth, Keswick and Penrith line did traverse the area towards its northern frontier, passing close to Blencathra and skirting the edge of Bassenthwaite. This apart, branch lines from outside the future National Park petered out after a few miles, as investors were deterred from further extensions by high building and maintenance costs, highly seasonal passenger traffic and limited prospects for sustained revenue from freight and minerals. Proposals might arouse fierce opposition, but this was never tested under circumstances of obvious commercial viability. The terminals, especially Windermere (and Keswick, as the most important Lakeland station on the one through route), became hubs for mid- and late-Victorian stage-coach and carriage traffic, as well as honey-pots for excursionists, while Lakeside, at the bottom end of Windermere, provided connections between a Furness Railway branch line and the lake steamers. But none of the railways were generally pro-active in encouraging popular tourism by excursion train: they preferred to focus on the safe, regular, high-class tourist trade which also patronised their hotels at Windermere and Keswick, and viewed trippers with a measure of suspicion.[21] The main exception was the Furness Railway at the turn of the nineteenth and twentieth century, especially under the enterprising management of Alfred Aslett, who from his appointment in 1896 was keen to develop tourist passenger traffic as the output of the region's mines began to falter. His full-colour posters, guide-books and promotions bore Edwardian fruit, especially when steamers from Fleetwood brought Blackpool holiday-makers across Morecambe Bay to Barrow to join circular tours of the Lake District by coach, train and lake steamer, often taking in Furness Abbey on the way, as Chapter 11 reminds us. Here was a further reminder that Blackpool and the Lake District were not mutually exclusive, and it was to be reinforced by a growing volume of charabanc trips from Blackpool and

19   W. Mudford, *Augustus and Mary, or the Maid of Buttermere* (London: M. Jones, 1803); M. Bragg, *The Maid of Buttermere* (London: Hodder & Stoughton, 1987).

20   H. Martineau, *Guide to the English Lakes* (Windermere and London, 1855).

21   J.K. Walton, 'The Windermere tourist trade in the age of the railway, 1847–1912', in O.M. Westall (ed.), *Windermere in the Nineteenth Century* (Lancaster: Centre for North-West Regional Studies, 1976); J.K. Walton and P.R. McGloin, 'The tourist trade in Victorian Lakeland', *Northern History* 17 (1981), pp. 153–82.

Morecambe in the inter-war years. This initiative failed to revive after the First World War, although combined rail and lake steamer excursions by other routes survived until the 1950s.[22]

Nor were railways necessarily emblems of modern efficiency, even in their ostensible heyday: regular train services might be slow and unreliable, especially on the line through Keswick, where weight restrictions left the traffic in the hands of ancient goods engines until the late 1930s, and the passenger carriages were 'antiquated, uncomfortable (and) dirty'.[23] Significantly, they came closest to matching the positive image when providing holiday expresses and luxury commuter services for middle-class visitors and wealthy residents, although there was nothing peculiar to the Lake District about such patterns of provision, which had their counterparts between Manchester and Llandudno, Southport and even Blackpool, while similar 'club train' services linked Leeds and Bradford with Morecambe from the late nineteenth century onwards, all disappearing in the 1960s.[24] Freight, the bread and butter of most Victorian railways, flattered to deceive in the Lake District, as Coniston's copper (especially) and slate production went into decline soon after the Furness Railway's branch line arrived there, while the heavy coke and iron ore traffic through Keswick had already peaked before the First World War.[25]

Almost all of the railways within the Lake District were closed between 1958 and 1972.[26] Apart from the remaining shuttle service on the Windermere branch, which has long lost the capacity to welcome excursions, and a brief incursion into the National Park by the surviving coastal route, Lakeland's surviving (or revived) railways have become tourist attractions in their own right. The Ravenglass and Eskdale line was, indeed, a pioneer in this respect. Opened in 1875 as a narrow-gauge line to bring haematite iron ore to the coast from Boot, it lost its original *raison d'être* with the closing of the mine in 1882, and staggered on by catering for tourists until closure in 1908. It then reopened, implausibly, as a miniature railway during the First World War, and linked up with charabancs to provide tours to Wastwater. After further vicissitudes it stabilised its fortunes under the auspices of a preservation society in 1961 and continues to flourish as a (mainly) steam railway in this guise. It was joined in 1973 by a more conventional contributor to the rapidly growing steam railway preservation industry, as part of the branch

22   D. Joy, *The Lake Counties: the Regional History of the Railways of Great Britain, Vol. 14* (Newton Abbot: David and Charles, 1983), pp. 116, 228–9; C. O'Neill, 'Visions of Lakeland: Tourism, Preservation and the Development of the Lake District, 1919–1939', PhD thesis, Lancaster University, 2000.

23   Joy, *The Lake Counties*, pp. 195–222, 227–31.

24   Ibid., p. 228; O.M. Westall, 'The retreat to Arcadia: Windermere as a select residential resort in the late nineteenth century', in Westall (ed.), *Windermere in the Nineteenth Century* (Lancaster: Centre for North-West Regional Studies, 1976).

25   Joy, *The Lake Counties*, pp. 206, 212.

26   Marshall and Walton, *Lake Counties*, p. 232; Joy, *The Lake Counties*, p. 215.

line to Lakeside reopened and restored a link with the sightseeing boat services on Windermere.[27] It is interesting to speculate on whether the fiercely contested railway proposals for the central Lakes in the late nineteenth and early twentieth century would, if built, subsequently have become candidates for preservation, as attitudes to railways shifted in response to the greater and more pervasive threat to peace and seclusion presented by road traffic. The revival of Welsh narrow-gauge lines for tourist purposes may not be a fair comparator: resistance to railways was always much stronger in the Lake District than Snowdonia, perhaps reflecting the former's special status as perceived literary landscape and epitome of a version of Englishness. It is, significantly, impossible to imagine a mountain railway ascending Helvellyn.[28]

Mining and quarrying, especially for copper, lead and slate, made an enduring impact on the Lakeland landscape, but not to the same extent as they did beyond its boundaries, on the Cumbrian coast or even through the granite workings to the east, around Shap. The copper mines on the side of Coniston Old Man finally gave up the ghost in 1942, after more than six centuries of working, and became an attraction for industrial archaeologists and, for a time, for venturesome or foolhardy underground explorers. Nearby slate workings also attracted their devotees, while the charcoal iron furnaces of Furness, using local ores and fed by distinctive landscapes of coppice woodland, extended into the southern Lake District. The Backbarrow furnace survived into the early twentieth century, and these activities also became magnets for industrial archaeologists after their working lives had ended.[29] Further north the Greenside lead and silver mine above Glenridding was a notorious polluter of the pristine waters of Ullswater until its closure in 1961; but many of its buildings have subsequently been adapted to Lake District recreational uses.[30] There were many other mineral and slate extraction sites scattered through the Lake District, including the Keswick mines for which German specialist workers had been imported in 1564, and the last surviving slate mining site at Honister in Borrowdale, where in 1883 the Lake District Defence Society had been formed to fight a proposed rail link, and where adventure tourism facilities by 2011 included a Via Ferrata. In the early 1990s the high and exposed Force Crag

---

27   Joy, *The Lake Counties*, pp. 218–22; H.I. Quayle and S.C. Jenkins, *Lakeside and Haverthwaite Railway* (Clapham, North Yorkshire: Dalesman, 1977).

28   I. Carter, *Railways and Culture in Britain* (Manchester: Manchester University Press, 2001); idem., *British Railway Enthusiasm* (Manchester: Manchester University Press, 2008).

29   J.D. Marshall and M. Davies-Shiel, *The Industrial Archaeology of the Lake Counties* (Beckermet: Michael Moon, 1977); E. Holland, *Coniston Copper* (Milnthorpe: Cicerone Press, 1987); M. Bowden (ed.), *Furness Iron: The Physical Remains of the Iron Industry and Related Woodland Industries of Furness and Southern Lakeland* (Swindon: English Heritage, 2000).

30   M. Hyde and N. Pevsner, *The Buildings of England: Cumbria* (New Haven and London: Yale University Press, 2010), pp. 366–7; I. Tyler, *Greenside and the Mines of the Ullswater Valley* (Keswick: Blue Rock, 2001); O'Neill, 'Visions of Lakeland'.

site in Coledale, near Braithwaite, which had produced lead, barytes and zinc, finally closed. Like the railways, all these activities attracted their enthusiasts, who were pulled together in the early twenty-first century by the Mines of Lakeland Exploration Society, while Keswick sustained its own mining museum.[31] There had long been an alternative, or additional, aesthetic of the mining and quarrying sublime, which was brought into combination with the extreme sports of mine exploration and climbing in these environments in the late twentieth century, and attracted the recognition and support of the National Trust.

These were not the only Lake District industries, as the industrial archaeologists were keen to point out. The mines and quarries provided a market for gunpowder in the eighteenth and nineteenth century, which in turn linked the area with international trade in additional ways, through imports of saltpetre (from India, China or Germany) and sulphur (from Italy) through local ports. The gunpowder mills tended to cluster around the fringes of Lakeland, but the Elterwater site, near Langdale, was right in the centre, close to the Langdale Pikes, and a holiday village was developed on the site in the 1930s, after its industrial life was over.[32] Increasingly, however, the dominant industry was tourism, with its associated service providers.

By its nature, the economic and social impact of tourism is never easy to establish on a statistical basis: indeed, the official figures remain subjective, contradictory and fundamentally untrustworthy, from those presented under World Tourism Organisation auspices to the most local of estimates. In this volatile, seasonal and fluctuating cluster of activities, with high levels of migration, tax evasion, multiple occupations and informal economies in association with large numbers of small businesses, such problems are only to be expected. They are compounded by the lack of satisfactory definitions of tourism and tourists, and the impossibility of counting and classifying convincingly. This has led to a proliferation of models based on heroic assumptions, whose implausible and ungrounded claims are often retailed as gospel by local authorities and tourist boards.[33] In a setting like the Lake District, where tourism has depended increasingly on private transport (whose traffic flows are much more difficult to quantify than those of large-scale public carriers) since the early twentieth century, informal activities which do not generate direct commercial footprints are highly important to the collective tourist experience, and small businesses which contribute to complex family economies are prevalent (especially where agriculture and tourism intersect), these problems are compounded. They were already apparent in Victorian Lakeland, when different sources gave contradictory readings of tourism's (highly seasonal) contribution to the economies of the three main urban centres of Keswick, Ambleside and

---

31   I. Tyler, *Honister Slate* (Caldbeck: Blue Rock, 1994); idem., *Force Crag: the History of a Lakeland Mine* (Marton: Red Earth, 1990).

32   I. Tyler, *The Gunpowder Mills of Cumbria* (Keswick: Blue Rock, 2002).

33   J.K. Walton and P. Browne (eds), *Coastal Regeneration in English Resorts – 2010* (Lincoln: Coastal Communities Alliance, 2010), chapter 8.

Windermere.[34] Margaret Capstick, analysing the economic prospects for North Westmorland in the late 1960s, was sceptical about the potential benefits of tourism for upland farmers to the east of the Lake District, and noted that a specialist Farm Holiday Guide for 1967 listed only 20 farmhouses in the whole of the 'Westmorland Lake District', perhaps suggesting that most did not advertise and depended on passing or casual bed and breakfast trade, thereby remaining invisible to most evaluative assessment. She did also observe that, 'Farmers in the Lake District dale heads, where the climbing fraternity congregate at least at weekends for most of the year, may find tourism income invaluable (evaluated at, in one case, 300 extra ewes).' A study of Dunnerdale, in the far west of the Lake District, found tourist income strongly biased towards the larger farms. But such fragmented evidence, with very limited potential for quantification (as the equation between climbers and ewes makes clear), was characteristic. Second homes and holiday cottages were making headway in North Westmorland; but here the problem of how to sustain the upland rural economy was already evident, especially if hill farm subsidies were removed: 'The hills could, of course, become derelict, their farmsteads abandoned or used only by holidaymakers, their pastures given over to bracken or rushes, their walls down and drains stopped. This form of dereliction is considered beautiful by many visitors to the countryside.' But, for Capstick, tourism was not the solution to rural economic problems: it was 'best considered as an independent industry with its own skills and opportunities.'[35]

The apocalyptic vision of a Lake District landscape no longer sustained by its small farmers, and either transmuting into a picturesque 'dereliction' which is no longer compatible with the Wordsworthian aesthetic, or being kept going artificially on a life-support system, haunts policy-makers in the National Park, as Susan Denyer reminded us in Chapter 1. Moreover, by the early twenty-first century falling incomes, threats to subsidies and the impact of official responses to the 2001 foot and mouth disease outbreak had brought about a real sense of crisis, which in turn generated vigorous attempts to promote recovery on the part of hill farming organisations. At one stage the foot and mouth disease episode threatened the slaughter of the Herdwick sheep flocks, and therefore the destruction of the whole upland management system.[36]

But by the early twenty-first century a complementary rural Lake District life support system was already in operation; and in contrast with Capstick's earlier

---

34   Walton and McGloin, 'Tourist trade'; J.K. Walton and P.R. McGloin, 'Holiday resorts and their visitors: some sources for the local historian', *Local Historian* 13 (1979), pp. 323–31.

35   Capstick, *North Westmorland*, pp. 120, 126–8, 130.

36   A.H.Griffin, 'If they go, it is the end of Lakeland', *Guardian* (11 April 2001); Voluntary Action Cumbria, 'Hill farming systems project Cumbrian fells and dales', http://www.cumbriahillfarming.org.uk/pdfs/areabasedevelopmentplan.pdf (accessed 1 June 2011). For Griffin see the obituary by J. Perrin, *Guardian* (12 July 2004), which also lists his many books on outdoor life in the Lake District.

views on North Westmorland, its dominant element was undoubtedly tourism; but in this case emphatically in a kind of symbiosis with agriculture and the desired landscapes it sustained. This was already demonstrated by official National Park figures for the difficult year of 2001. Specialised hotels and guest-houses accounted for only 20 per cent of recorded staying visitors; and they would be more likely to be fully represented in the statistics than more informal accommodation providers. So would the Youth Hostels which accounted for 4 per cent of recorded visitors. 28 per cent stayed in self-catering cottages, a thriving industry in its own right, with some direct links to farm economies. Farmers would also provide for many of the 8 per cent of visitors who stayed on a hundred sites for static caravans, and the further 8 per cent who pitched their tents on official camp-sites. Most of the 14 per cent who took bed and breakfast accommodation, but not in a guest-house, were probably staying on farms. Examination of these statistics illustrates the ways in which tourism was permeating the rural Lake District economy, but also the problems inherent in attempting to disentangle it from other activities.[37]

The only positive aspect of the foot and mouth disease crisis was the way in which it stimulated renewed efforts to stimulate and diversity rural tourism provision as a complement to agriculture. The following decade saw an accelerating tendency for farms to diversify more systematically into tourism, with (for example) the conversion of redundant barns into luxury accommodation (in perhaps disturbing contrast with earlier bunkhouse provision), and the eventual conversion of tourism into the main source of rural income in South Cumbria.[38] This was not a uniform process, and it coexisted with efforts to sustain, in their own right, the farming lives and practices which in turn sustained the valued landscapes and, in turn (although in context this was of lesser importance) the World Heritage Site bid.

This rural tourism development was the end product of a long and unobtrusive history, resistant to conventional economic analysis, which had begun before the transfer of most incoming tourist traffic to the roads and the private motorist. This was a long process which made its own insidious impact on landscape and society over a century, and reached its tipping point in the late 1950s and early 1960s, whether the preferred destination was coast, countryside or (by then) National Park. Under the circumstances, visitor numbers in the Lake District in the mid-1960s (when nearly half of British households had a car) were not, at first sight, overwhelming. On August Bank Holiday Sunday, 1966, 10,000 visitors arrived at Grasmere, 6,000 at Coniston Water, 5,000 each at Derwentwater and Borrowdale, and 2,000 in Langdale. Many of these stayed only for a few minutes as part of a long tour, usually along main roads, so the numbers at any one time would be much lower. The three main towns were much more popular as destinations, with between 15,800 and 19,500 visitors. As John Marshall has pointed out, the Lake District's visitor numbers have never rivalled those of so-called 'mass

---

37   http://www.lakedistrict.gov.uk/g_tourism.pdf (accessed 2 June 2011).
38   R. Ryan, 'Tourism "wake-up call" in Lake District after foot-and-mouth', *Cumbria News* (15 February 2011).

tourism' destinations in other parts of England: in 1961, for example, there were 520 hotels and boarding-houses in the whole of Cumberland and Westmorland, compared with 880 (at that time) in Morecambe, and over 4000 in Blackpool alone. Maximum visitor numbers for the whole region in the early 1960s were around 66,000, and perhaps two-thirds of a million holidays were spent in the two old counties in 1960.[39] But the temporary impact of even these numbers, especially when translated into cars in the rural landscape, was nevertheless considerable. Their numbers were 'greatly in excess of the capacity of recognised parking areas and laybys', and pressures were mounting on policy-makers to restrain demand. The role of the private car was already highly significant in the evolution of rural tourism, as its flexibility enabled tourist penetration of the countryside, and put pressure on conventional versions of rural amenity, to a completely novel extent and in a far more generalised way than the contentious activities of power-boat enthusiasts on particular popular lakes.[40] The subsequent growth of car-based tourism has led to sustained pressure for road improvements, and to great expansion in car parking facilities, with appropriate screening by trees in sensitive locations. But, as noted above, the impact of the car has been channelled and limited by the ways in which most people use it. The dominant preference for main roads, beaten tracks and honey-pots has been thoroughly documented by researchers, as has the tendency of most visitors, even to the Lake District, not to move far from cars parked in convenient locations, preferably close to water. This has made it easier to sustain the quieter aspects of the Lake District experience in ways which are consistent with the Wordsworthian agenda and the agriculture which sustains it. The great post-war expansion of the landholdings of the National Trust, which now owns a quarter of the National Park, has been very significant. The Trust's countryside management policies, together with those of the National Park (with its highly popular visitor centre at Brockhole), have helped to further this.[41]

In many ways the vulnerability of the rural Lake District economy at the beginning of the new millennium, and the complex relationships between farming and tourism, were brought into sharp relief by the foot and mouth disease crisis of 2001. This was, as the veterinary historian Abigail Woods has ably demonstrated, a 'manufactured plague'. It had been thus constructed by officialdom since the late nineteenth century, and a minor inconvenience to most of the animals affected had been converted, rhetorically and administratively, into a dangerous monster requiring the extermination of flocks and herds on a Biblical scale. The slaughter in Cumbria was particularly widespread, affecting and damaging to farming livelihoods, even though the Herdwicks of

---

39   Marshall and Walton, *Lake Counties*, pp. 234–5.

40   J.A. Patmore, *Land and Leisure* (Harmondsworth: Penguin, 1972), pp. 113–18, 139.

41   For the changing content of tourism policy in the Lake District at the beginning of the twenty-first century see D.W.G. Hind and J.P. Mitchell (eds), *Sustainable Tourism in the English Lake District* (Sunderland: Business Education Publishers, 2004), chapters 4–8.

the fells were eventually spared.[42] The official response, which incorporated the completely unscientific pretence (in the sense that there was no supporting evidence) that foot and mouth disease could be spread by human pedestrians, compounded the damage through indiscriminate footpath closures and more general restrictions on movement which effectively made most aspects of Lake District tourism untenable for most of a summer season, not least because farmers understandably believed the pretence and regarded walkers as serious threats to their livestock and livelihood. The impact was devastating, with job losses running at 350 per week at one point, a 6.6 per cent loss in visitor revenue for the year, a fall in tourism trips to Cumbria from 2.3 million to 1.8 million, a fall in hotel employment from 9633 full-time equivalents to 6843 between 2000 and 2001, and a large number of attractions having to close for three months, many of which never reopened. In the middle of all this the Ambleside and Keswick tourist offices illustrated the absurdity of the footpath closure policy by reminding potential visitors that 'you can still walk on the roads', many of which were unfenced and open to livestock, and arranging and advertising walking and cycling routes.[43]

In the wake of the foot and mouth crisis, tourism has at last become identifiably the most important economic activity over much of the Lake District, especially in the southern half of the National Park. As John Marshall pointed out at the time, this was emphatically not the case in the early 1980s, and it was probably still not so at the turn of the millennium, although allowance should be made for the subtle and nuanced ways in which tourism contributes to economic development and social change.[44] The recent nature of this transition must be emphasised, not least because it has quickly come to be taken for granted. A recent American article, in assuming the dominance of tourism, provides a table presenting the employment of the resident population of the National Park in 2006 which, as the author blithely admits in passing, has no categories for agriculture or forestry, which are assumed to have become residual or irrelevant.[45] After all, it now seems to be the case that all of the Lake District's old industries, as well as agriculture itself,

---

42   A. Woods, *A Manufactured Plague: the History of Foot-and-Mouth Disease in Britain* (London: Earthscan, 2004); A. Woods, 'Why slaughter? The cultural dimensions of Britain's foot and mouth disease control policy, 1892–2001', *Journal of Agricultural and Environmental Ethics* 17 (2004), pp. 4–5.

43   R. Allison, '1,000 miles of Lake District footpaths reopened', *Guardian* (2 August 2001); E. Hartley, 'Tourism staff protest at job losses', *Independent* (22 March 2001); W. Irvine and A.R. Anderson, 'The impact of foot and mouth disease on a peripheral tourism area', https://openair.rgu.ac.uk/bitstream/10059/214/1/Anderson17.pdf (accessed 2 June 2011); and see above, note 36.

44   Marshall and Walton, *Lake Counties*, p. 235.

45   A.J. Scott, 'The cultural economy of landscape and prospects for peripheral development in the twenty-first century: the case of the English Lake District', *European Planning Studies* 18 (2010), pp. 1575–6. The table will also *understate* the importance of tourism by omitting seasonal migrant workers.

have recycled themselves as attractions for tourists. This even applies to the most attractive of the industrialists' lakeside mansions of the late nineteenth and early twentieth century, especially their landscaped gardens. And this is a reminder that the landscape itself has become, perhaps above all, an attraction in its own right whose economic justification (leaving others aside) might be represented in terms of the jobs it generates and the tourism revenue it provides. Here is, perhaps, irony piled on irony, with the World Heritage Site proposal adding its own extra layers of paradox and contradiction. Here is post-industrial post-tourism; but here is also a working landscape which sustains its own enduring values in symbiosis with the literary and philosophical traditions with which it has become indelibly identified, and which in turn pulls in the tourists.

Current official perceptions of the nature of and prospects for tourism in the Lake District are revealing.[46] Although the long-established tendency for statistical assessments to understate the economic contribution of tourism is still apparent, and important aspects of tourism businesses remain below the radar, Cumbria Tourism believes that the picture presented by the widely-used STEAM modelling system is more trustworthy for its territory than may be the case elsewhere, although (for example) its data on day-trippers up to 2009 did not inspire confidence.[47] Much of the available data covers the county of Cumbria, or administrative districts within it, rather than the National Park, which at the census of 2001 contained one-third of the county's area but only 41,650 of its population of 487,608.[48] This is important because the visitor numbers for the county as a whole (calculated at 15.3 million in 2008) are said to have amounted to 7 million more than those to the National Park.[49] We do not know how many visitors spent time in both the National Park and the 'rest of Cumbria', though they would have to travel through the 'rest of Cumbria' to reach the National Park; but in any case the characteristics of visitors to 'Cumbria', and especially to the outer concentric rim, may have been very different from those of the substantial sub-set who made their way into the Lake District 'proper'. The psychographic breakdown of visitors interviewed in the 2006 Cumbria Visitor Survey suggested that Cumbria's 'core market' of 'traditionals' and 'functionals' was actually most strongly identified with Hadrian's Wall and Carlisle, and to a lesser extent with the Settle and Carlisle Railway and Keswick (which was, for some reason, abstracted from 'the Lake District'). The Lake District as a category was not especially attractive to this

---

46   Richard Greenwood of Cumbria Tourism, who was prevented from contributing a chapter to this volume by the pressure of coping with savage cuts to his organisation's budget, kindly provided much of the material for this section. He is not responsible for the uses to which it has been put.

47   Locum Consulting, *Cumbria Market Forecasts: 2007 Update* (Haywards Heath: Locum Consulting, 2007), p. 71, note 48; information from Richard Greenwood; and see above, note 33.

48   Scott, 'Cultural economy', p. 1575, table 1.

49   Scott, 'Cultural economy', p. 1576.

'core market', although it did quite well among the 'style hounds' group. The attempt to analyse visitor markets by psychographic characteristics is interesting, but some reservations must apply to a classification system whose 'cosmopolitans' are most strongly drawn to Stockport and the Botany Bay retail outlet near Chorley, while the most favoured destinations of the 'high street segment' were St Helens, Blackburn and Burnley. Nor is it clear how individual interviewees were assigned to categories. There is, however, no doubt that subsuming the National Park into a 'Greater Cumbria' is likely to dilute and even distort representations of its own distinctive visitor profile.[50]

There does seem little doubt that some of the 'key findings' of the 2009 Visitor Survey for Cumbria will be valid in outline for the Lake District, as they are congruent with other evidence over a long period. Visitor numbers seem to be holding up well (a 'gentle increase' since the temporary fall in 2001), without expanding dramatically, although the similarities between recent STEAM figures year on year, category by category, may provoke a measure of suspicion. Only 7 per cent of all visitors came from outside the United Kingdom, and most of those sampled were repeat visitors from northern England: 25 per cent even of the day-trippers came from the North-West. But 'almost half of all overseas visitors were also on a return visit'; and only 15 per cent of the total were first-timers. Nearly two-thirds of the visitors were over 45 years old, and only 15 per cent were under 34. So this was an ageing and apparently conservative visiting public, with a predominantly regional catchment area, which knew what to expect and was happy to return for more.[51] In many ways this profile was similar to that of Blackpool's first visitor survey in 1972, which prompted alarm about the apparent failure to recruit the rising generation. An important contrast was Cumbria's predominant ABC1 (professional, managerial, white-collar) profile, which contrasted with that of the proletarian popular resort, and justified greater optimism.[52] The other big difference was the overseas presence among visitors to Cumbria; and the 2007 market forecasts study suggested that this was beginning to diversify beyond the traditional Anglophone market of the United States, Canada and Australia, which accounted for just over one-third of the overseas visitors in the sample. Germany now accounted for 12 per cent of overseas visitors (but still less than one per cent of the whole), and five other Western European countries made up 20 per cent, with the Japanese contingent contributing 5 per cent.[53] These proportions were probably higher for the National Park itself; but despite these modest changes, there is nothing to suggest that the Lake District enjoyed high visibility or reputation

---

50   Locum Consulting, *Cumbria Market Forecasts*, pp. 81–96.

51   H. Tate, *Cumbria Tourism Visitor Survey 2009* (Cumbria Tourism summary document, 22 February 2010, by courtesy of Richard Greenwood).

52   J.K. Walton, *The British Seaside: Holidays and Resorts in the Twentieth Century* (Manchester: Manchester University Press, 2000), p. 66; Locum Consulting, *Cumbria Market Forecasts*, p. 61.

53   Locum Consulting, *Cumbria Market Forecasts*, p. 63.

beyond those parts (and social strata) of the English speaking world that shared the cultural capital of the Picturesque aesthetic and the Romantic sensibility.

It is interesting, indeed, that Cumbria Tourism and its consultants tend to take landscape aesthetics and literary associations 'as read', and to seek the further expansion of tourism by other routes. The main thrust of their proposals for increasing visitor numbers, and especially per capita visitor spend, seems to be identified with those parts of Cumbria that are outside the National Park. The plan for a 'West Lakes Extreme' attraction, based outside the National Park on the former mining town of Cleator Moor, and offering indoor canyoning and caving as well as a snowdome, remains no more than a gleam in entrepreneurial eyes; but it fits into a pattern of new attractions accumulating beyond the fringes of the National Park, at the stately homes of Levens, Holker and especially (under ambitious new management) Lowther Castle, while attention is also being directed towards Carlisle and the coastal districts. Within the National Park there is interest in heritage, culture and events, including street theatre, culture and drama, with aspirations towards establishing an iconic new museum or gallery on the run-down Bowness waterfront, close to a renewed Steamboat Museum, and (more immediately) to the development of (non-mechanised) water sports on Windermere.[54] The Locum Consulting market forecasts document recognises the existence of a category of 'old scenery watchers', who are 'relatively inactive', and low spenders but frequent visitors; but there is no category for walkers, climbers or what might be called active contemplatives.[55] The central thrust of the 'planning for growth' agenda is directed towards improvements in accommodation and catering standards. The latter entails the encouragement of locally sourced ingredients and regional recipes, thereby providing a recognised role for local agriculture in symbiosis with tourism; but there is no active engagement with the 'traditional' agenda of a Lake District visit. Such activities as walking, climbing, contemplating 'Nature and Nature's God', encountering (an idealised) rural culture and appreciating a sacralised literary landscape are perhaps not sufficiently integrated into market-based assumptions of economic activity to qualify as 'tourism' for planning purposes. They are just 'there'.

We should therefore not be surprised that Cumbria Tourism and its predecessor bodies have been resistant to, or at best equivocal about, the World Heritage Site proposal. When, in 2005, Cumbria County Council was consulting 'local community partners' about whether it should support the current bid, Cumbria Tourism (then the Cumbria Tourist Board) responded that WHS status was 'likely to have a dampening effect on the economy, regeneration, housing, investment and tourism since this is primarily a preservation/conservation measure'. It would 'add to existing planning restrictions thus discouraging investment and regeneration' in an area which was already well catered for in this respect, and 'might assist in making the case for reducing car traffic in certain areas', with a possible 'negative

---

54   Information kindly supplied by Richard Greenwood.
55   Locum Consulting, *Cumbria Market Forecasts*, p. 79.

impact on some visitor dependent businesses, and reduce the enjoyment of car bourne (*sic*) visitors.' 'The cultural benefits appear to be limited in scope focussing mainly on landscape and the heritage/historical associations with art and literature.' These priorities and assumptions have been faithfully reflected in the development of Cumbria Tourism's plans and policies.[56]

The actual World Heritage Site proposal has indeed emphasised landscape, literature and literary heritage, on a broad canvas within its own framework of assumptions, which focus on what has made the Lake District special and of 'Outstanding Universal Value'. As we have seen, it does not confine itself to the artistic and literary dimensions, but also highlights the need to sustain the distinctive agricultural system, together with the values that are held to underpin it historically. This in turn sustains landscape and society, preserved but not in aspic, indeed in moderated evolution. Furthermore, the influence of the Lake District and its resident intellectuals as an incubator for conservationist thinking, and its importance as the originator and springboard for the distinctive British model of National Park designation (now held to be gaining ground against older and more exclusive models), have been incorporated into the bid.[57] More recently, the Lake District World Heritage Steering Group has commissioned a substantial piece of research on the potential opportunities for 'economic gain' from World Heritage Site status, which identifies a growing trend for bids and designations to be associated with 'branding' and 'place making' activities in competitive international markets, and thereby intersects more closely with the concerns of Cumbria Tourism, as well as those of the project's sponsors, the North West Regional Development Agency and 'Invest in Cumbria'.[58] What Wordsworth or Ruskin would have thought of this is another question, located, perhaps, in a different moral universe.

---

56   Cumbria Tourist Board, response to Cumbria County Council consultation on World Heritage Site bid, reply to message from Rebecca Hughes dated 21 July 2005, by courtesy of Richard Greenwood.

57   Chris Blandford Associates, *Lake District Candidate World Heritage Site: Outline Statement of Outstanding Universal Value* (London: Chris Blandford Associates, October 2006), section 2.

58   Rebanks Consulting Ltd and Trends Business Research Ltd, *World Heritage Status: Is There Opportunity for Economic Gain?*, research commissioned 2009.

Chapter 5

American Tourists in Wordsworthshire: From 'National Property' to 'National Park'

Melanie Hall

A series of influential Americans visited the Lake District during the nineteenth century predominantly because of Wordsworth and his writings.[1] Wordsworth was pre-eminently known as the poet of rustic nature in a century when important debates focused around nature's representation. Part celebrity, part social critic, his concerns with society and with the reach of English literature gave Wordsworth an interest in the development of the newly independent United States which made him attractive to Americans. His emotionally evocative, site- and excursion-based poetry also referenced travel from modernising landscapes to 'cultural homelands' (to use Leonard Tennenhouse's phrase for literary pastoral landscapes) in which, as Leo Marx observes, 'the theme of withdrawal from society into an idealized pastoral landscape' provided an imaginative 'middle ground' where extraordinary changes could be mediated for readers at home and abroad.[2] For Wordsworth, the 'cultural homeland' and the 'middle ground' were at once a literary art form and an actual landscape. In his *Guide to the Lakes* (1810, re-issued more popularly in 1835), he suggested that discerning tourists had already deemed the region a 'sort of national property'; later, this was perceived as prescient and read as a call for the landscape of his inspiration to be made a public property.[3] For Americans, both Wordsworth's poetry and the landscape that inspired him offered points of reference and comparison as they sought to turn literary and artistic 'middle grounds' into national properties. I argue in this preliminary discussion of a complex phenomenon that, at the turn of the twentieth century, two of Wordsworth's

1   I am grateful to Erik Goldstein, Bernard Richards, Alan Trachtenberg, David Thomason and John K. Walton for their helpful comments, suggestions and insights as I developed this chapter. I am also grateful to Anita Israel, archivist at the Longfellow House, and to Jeff Cowton at the Centre for the Study of British Romanticism, Grasmere for invaluable assistance with sources and documents in their care. For permission to quote from his archive, I am grateful to Alexander Yale Goriansky. Part of this research was made possible by a grant from the Friends of the Longfellow House, Cambridge, Massachusetts.
2   L. Tennenhouse, *The Importance of Feeling English* (Princeton, NJ: Princeton University Press, 2007), esp. pp.17–18; L. Marx, *The Machine in the Garden, technology and the pastoral ideal in America* (New York: Oxford University Press, 1964, 2000), p. 10.
3   *Guide to the Lakes. William Wordsworth*, ed. by Ernest de Sélincourt, with a preface by Stephen Gill (London: Francis Lincoln, 2004), p. 93.

imaginative, literary 'middle grounds' were realised as sort of 'national parks' by advocates seeking to reserve representational sites for recreational leisure pursuits. Brandlehow and Gowbarrow, the first such landscape sites in the Lake District were set aside as 'national property' in 1902 and 1904 by the National Trust for Places of Historic Interest and Natural Beauty (as the organisation was known when founded, 1894–95), and Anglo-American literary relationships were key to this development.

Landscapes of Romantic Nationalism

'Nature', as Peter Harman and Simon Schama among others demonstrate, is a cultural landscape represented in Arcadian, pastoral, theological or 'spiritual' metaphors as well as more philosophical and political constructs.[4] During the nineteenth century it was increasingly redefined in rational, scientific, geographical, geological, individual and social terms in the culturally European world. A range of important debates was organised around the use and stewardship of 'nature's resources'; Wordsworth joined these debates.[5] American writers, too, focused on nature as the vast continent was settled and explored. Tennenhouse describes their task in the years of the colonial period and early republic as evoking an imaginative 'cultural homeland', not quite English, but a new and adapted version for a diaspora.[6] Wordsworth was similarly engaged in representing a cultural homeland, depicting the Lake District as a place at once real and imaginative. His readers at home were navigating a changing, revolutionary landscape as were those English-speaking readers who still included Americans in whose national project he took interest. In 'The Brothers' (1799), Wordsworth depicts change in the Lake District by juxtaposing religious practices with modern tourist practices; his 'homely Priest of Ennerdale' intones, from 'Tourists, heaven preserve us!'.[7] Wordsworth describes different kinds of travellers; here, the 'tourist' returns to his homeland but, changed by experience, his brother dead, he does not stay. In the poem, Wordsworth creates a 'middle ground' between the priest's old, religious world and the brother's modern, distant world; simultaneously, he makes poetry a mediator of that change, and situates himself as a modern 'cleric' (discussed below). In alluding to the Lake District as a 'sort of national property', Wordsworth sought to reserve the area for aesthetic memory-

---

4   P.M. Harman, *The Culture of Nature in Britain, 1680–1860* (New Haven and London: Yale University Press, 2009); S. Schama, *Landscape and Memory* (New York: Alfred A. Knopf, 1995).

5   N. Everett, *The Tory View of Landscape* (New Haven and London: Yale University Press, 1994), esp. pp. 204–7.

6   Tennenhouse, *Importance of Feeling English*, esp. pp. 17–18.

7   W. Wordsworth, *The Poems*, ed. J.O. Hayden (New Haven: Yale University Press, 1981), vol. 1, p. 402, quoted in J. Buzard, *The Beaten Track: European tourism, literature, and the ways to culture, 1800–1918* (Oxford: Clarendon, 1993), pp. 1, 19–27.

journeys that his own poetry would help direct, and to represent the Lake District as he perceived it as a 'middle ground' or 'cultural homeland' poetically. As yet there was no means to realise this aspiration in reality.

The creation of imaginative literary homelands was noticeable in New England; critic Margaret Fuller, who promoted Wordsworth's poetry in the United States, saw in his contribution to England's national project a model for Americans.[8] In describing American writers as 'brother bards across the Atlantic' and expressing a desire to 'see both countries united more and more strongly in the bonds of brotherhood', Wordsworth provided a poetical bridge between the two literary cultures.[9] As a spokesman for poetry and society, Wordsworth was a point both of arrival for tourists and of departure for American literature. Aspects of the complex relationship played out in American accounts of Wordsworthshire as Americans, too, sought to reserve cultural landscapes from change. Arguably, at this cultural interface a new response to change was formed as 'cultural homelands' became realised as national parks.

Steam facilitated travel; it also altered the landscape and so the relationship between the author and the pastoral ideal, and between the reader – who might also be a traveller or tourist – and the landscape. In *The Machine in the Garden*, where the garden is the idealised prospect of the early American Republic, Leo Marx considers the development of pastoral literature. While this accorded with a pastoral view in politics espoused particularly by Thomas Jefferson, American literature in the Arcadian tradition was able to provide an imaginative 'middle ground' where change could be mediated.[10] As the machine, that is the railway, gradually altered this relationship, access to America's 'wilderness' beyond the garden was facilitated. From being something that could assist man in harnessing nature and, thus, enhance the landscape for man's benefit and 'progress', by the end of the century the machine seemed to some to be something that could usurp 'the wilderness'. As it did so, it disturbed conceptions of both the pastoral and the moral (Edenic) landscape that helped sustain the new republic's self-image.

As Marx states, retaining the pastoral myth in sentimental guise 'enabled the nation to continue defining its purpose as the pursuit of rural happiness while devoting itself to productivity, wealth, and power. It remained for our serious writers to discover the meaning inherent in the contradiction.'[11] By the 1880s, it was apparent that existing literary 'middle grounds' were no match for the machine. New industrial and economic demands on the landscape required new cultural

---

8   C. Capper, *Margaret Fuller, An American Romantic Life*: *the private years*, 2 vols (New York: Oxford University Press, 1992), vol. 1, p. 179.

9   'A.M.', *Southern Literary Messenger* (August 1850), vol. 16, pp. 474–9, relaying a conversation with the poet, quoted in K. Karbiener, 'Intimations of Imitation: Wordsworth, Whitman and the Emergence of *Leaves of Grass*', in J. Pace and M. Scott (eds), *Wordsworth in American Literary Culture* (Basingstoke: Palgrave Macmillan, 2005), p. 150.

10   Marx, *Machine in the Garden,* esp. pp. 3–18, 195–200, 356–65.

11   Ibid., p. 226.

responses not only to the landscape but also in the landscape; prescience, politics and pragmatism called for practical solutions. Sciences, travel and tourism, and the recognised benefits of outdoor exercise, required actual *and* imaginative 'middle grounds' that were both 'beautiful and useful' (to reference William Morris), as did politics. This was not an unequivocal issue; as machines took control of 'wilder' nature, more advanced utilitarian necessities were provided for expanding populations, at the same time 'wilderness', which had become important to the American identity, had to be set aside if it were to be reserved. A new genre of preserved sites associated with literature, art, 'wilderness' and other narratives developed as 'national' and state parks. And gradually, more developed pastoral landscapes also came to be set aside, as we shall see.

Designating and institutionalising 'cultural homelands' is not unique to the Anglo-American world; Sweden's 'fornhem' and German 'heimat' offer parallels.[12] In defining and reserving such landscapes of romantic nationalism, nations recognised one another's progressive and contemporary contributions to a new tradition of 'Western civilisation'. Thus, they had both national and inter-national characteristics.[13] This chapter opens up a discussion of a complex phenomenon. Its focus is the Lake District but it has, I think, implications for the making of other 'cultural homelands' throughout the English-speaking worlds.

Marx explains how Wordsworth juxtaposes the 'rash assault' of the railway in the landscape – the machine in the garden – with what he regarded as nature's tranquillity.[14] Railways radically altered the face of the landscape; they also altered man's performative and emotional engagement with the landscape, as J.M.W. Turner famously evoked in 'Rain, Steam, and Speed' (1844). John Ruskin understood Romantic art's ability to associate emotion with specific landscapes and thus, to memorialise them; Romantic poetry had a similar effect but memorialising an actual and changing landscape was problematical.[15] Wordsworth's Romantic poetry engaged with pastoral sites, sensory experience and 'natural elements' (weather conditions). He regarded walking excursions as the best mode of travel around the area. In nominating the Lake District a 'sort of national property' in which all those 'with an eye to perceive and a heart to enjoy' have an interest, he represented this area's 'natural' features and aspects of its way of life as of utility

---

12   O. Wetterberg, 'Conservation and the Professions: the Swedish Context 1880–1920', in M. Hall (ed.), *Towards World Heritage: International Origins of the Preservation Movement, 1870–1930* (Aldershot: Ashgate, 2011), pp. 201–20. C. Applegate, *A Nation of Provincials: the German idea of Heimat* (Berkeley: University of California Press, 1990).

13   See, for example, A. Swenson, 'The Law's Delay? Preservation Legislation in France, Germany and England, 1870–1914', in Hall, *Towards World Heritage*, pp. 139–54, and introduction by Hall in same volume.

14   Marx, *Machine in the Garden*, p. 18; W. Wordsworth, 'On the Projected Kendal and Windermere Railway', *The Poetical Works of William Wordsworth,* 7 vols (Boston: Little, Brown and Company, 1854), vol. 2, p. 395.

15   Harman, *Culture of Nature*, p. 148. Turner depicts the Falls in full flood, an infrequent scene.

to national society as a re-creative reserve.[16] This representation functioned on several levels, as we shall see.

Representational Culture

Soft-power diplomacy requires that government and the law represent their presence and associate that with attractive socio-cultural values. In a democracy (however limited or virtually represented it may be) successful representation must appeal proportionately to the electorate. Landscape for centuries had provided contexts, concepts and representation for land ownership and other rights. In appealing to ideals of 'nature', Wordsworth invoked metaphors that had underpinned society's fabric in complex legal and political ways.[17] 'Nature' was topical as science redefined it and new industries represented themselves in the landscape in competing ways. Railways were particularly contested; although facilitating travel they altered destination sites, sometimes compromising them with detritus and development. Modern tourism and, particularly, literary tourism can be understood in terms of invented traditions and imagined communities acted out in national landscapes; it demanded direct aesthetic and sensory experiences in specific, cultural arenas.[18] In his *Guide*, Wordsworth was not merely memorialising; he was engaging readers and tourists in his political and cultural ideals and aspirations. Wordsworth's *Guide* represents the poet as a member of a distinctive rural community with its own society and ideals of stewardship. He invokes a rustic-picturesque landscape through poetry (that is, mytho-poetically) and through the more rational and practical medium of a guidebook which, in turn, can be followed. In describing a cultural landscape in political terms and calling for its designation as a 'sort of national property', Wordsworth joined existing debates about national property and property rights. He represents Westmorland as having its own culture of stewardship maintained by the remains of an English yeomanry who cared for farmland and common land to the communal good.[19] Expressing their local

---

16 *Guide to the Lakes*, p. 93. Everett, *Tory View*, pp. 204–07.

17 Harman, *Culture of Nature*, pp. 152–64; G. Wood, *Empire of Liberty: a History of the Early Republic, 1789–1815* (Oxford: Oxford University Press, 2009), pp. 45–6, 117–18.

18 See E. Hobsbawm and T. Ranger (eds), *The Invention of Tradition* (Cambridge: Cambridge University Press, 1983); B. Anderson, *Imagined Communities: Reflections on the Spread and Origin of Nationalism* (London and New York: Verso, 2006), esp. pp. 10–13, 37, 44–5, 52–6, 67–9, 200–6; M. Kammen, *Mystic Chords of Memory: The Transformation of Tradition in American Culture* (New York: Alfred A. Knopf, 1991).

19 J.K. Walton, 'The strange decline of the Lakeland yeoman: some thoughts on sources, methods, and definitions', *Trans. Cumberland and Westmorland Antiq. and Archaeol. Soc.*, new ser., 86 (1986), pp. 221–33; J.D. Marshall, '"Statesmen" in Cumbria: the vicissitudes of an expression', *Trans. Cumberland and Westmorland Antiq. and Archaeol. Soc.* new ser., 72 (1972), pp. 249–83; B.L. Thompson, *The Lake District and the National Trust* (Kendal: Titus Wilson & Son, 1946), pp. 10–11, 15, 72–3.

building tradition as 'natural', he depicts a model rural landscape in expanded poetical, rustic picturesque terms.[20] Although philosophically conservative, this offers a contrast to the dominant assumptions of an aristocracy who had enclosed common land while laying out private, Arcadian pleasure gardens that invoke classical metaphors of ownership and leadership as the polite picturesque. By mythopoeticising a communal, rustic-picturesque landscape Wordsworth engaged broader, political debates about property rights, cultural leadership, 'the national community', rights of access and control of representation. These debates were topical in philosophy and in the developing public domain of state-supported museums, and would include Kew Gardens by the 1840s.[21]

In an influential essay, *On the Constitution of the Church and State* (first published 1829 and subsequently reprinted with revisions), Wordsworth's sometime fellow Lakeland poet and early collaborator Samuel Taylor Coleridge developed a model for the poet as a member of a modern '*enclesia*' or 'a permanent nationalized learned order'. This would be for the 'peace and weal' and 'happiness' of the nation, as Raymond Williams explains, by 'the cultivation of learning, and for diffusing its results among the community' in order that they will understand 'their rights, and ... duties correspondent'.[22] In representing Westmorland society and environment, Wordsworth adopts a professional, custodial and 'clerical' role (akin to the priest in 'The Brothers'). Artistic and narrative references already helped to define certain sites, such as Stratford-upon-Avon, and provided utility to a range of industries including tourism and, increasingly, education.

Although beyond the scope of this chapter to assess his impact on preservation thinking, it is Coleridge's idea of such a 'National Church' held as a 'National Trust' that provided the most significant underpinning philosophy for pastoral, preservationist thinking throughout the century and, surely, the name for the National Trust.[23] Tourism demanded realised sensory pleasures and cultural engagement. In associating site-seeing, rambling and excursions, and belief (as 'natural' theology) with socio-political ideals expressed in a particular landscape, Wordsworth uses rustic-

---

20   Everett, *Tory View*, pp. 159–85. For definitions of categories of picturesque see P. Goodchild, *Humphrey Repton, On the Spot at Mulgrave Castle* (Halifax: Stott Brothers, 1985). I am grateful to Peter Goodchild for discussions of the picturesque. See Harman, *Culture of Nature*, p. 96 for rustic tradition.

21   A. Goldgar, 'The British Museum and the virtual representation of culture', *Albion* 32, 2 (Summer 2000), pp. 195–231; R. Drayton, *Nature's Government: Science, Imperial Britain and the 'Improvement' of the World* (New Haven and London: Yale University Press, 2000), pp. 198–201, 215–17; Drayton observes these debates from the 1840s.

22   S.T. Coleridge, *On the Constitution of the Church and State, according to the idea of each; with aids towards a right judgment on the Catholic Bill* (London: Hurst, Chance and Co., 1830), pp. 46, 66, 70, 74–7, 81–2; R. Williams, *Culture and Society 1780–1950* (London: Chatto & Windus, 1958), esp. pp. 3, 29, 56–7, 62–4; Everett, *Tory View*, pp. 205–6; for its influence in New England see J. Turner, *The Liberal Education of Charles Eliot Norton* (Baltimore and London: The Johns Hopkins University Press, 1999), p. 12.

23   Coleridge, *Church and State*, pp. 57 et passim, and pp. 83, 84, 86, 123.

picturesque, pastoral, autobiographical, and patriotic metaphors. Simultaneously, he associates the site and its values with the value of his own work and status, and with 'the nation'. Further, in writing his *Guide* for those with 'an eye to perceive and a heart to enjoy', Wordsworth invokes Old and New Testament analogies to establish his place as a modern 'clerical' leader. Following his works and walks, place helps to anchor belief through emotion and experience; here, emotion can be split between an appreciation of Nature, an appreciation of Culture (poetry), an admiration for the poet as 'hero', and a more patriotic sense of place. Seen retrospectively, the cultural landscape of the Lake District, perceived as a 'sort of national property' by Wordsworth, might also become a sort of 'national reserve', an important and influential concept expressed by Coleridge as a type of spiritual and social fund for the 'cultivation' of the nation, and which was subsequently defined as 'culture' by John Stuart Mill.[24] Coleridge explained the reserve as 'the stores and ... treasures of past civilization, [that] thus ... bind the present with the past'; it was further the job of the intellectual elite 'to perfect and add to the same, and thus to connect the present to the future', while their followers were to preserve them.[25]

Wordsworth invokes emotional-moral support and ideas of well-being for an alternative model of national stewardship that was more suited to the gentry and rising middle class. The Lake District has 'a strikingly large proportion' of common land, despite extensive enclosure in the late eighteenth and early nineteenth century.[26] In further considering that this cultural homeland should be reserved as a 'national property', Wordsworth did not so much call for landscape protection as provide a locus for debate about communal property and its proper stewardship in the modern, public realm. These representational issues became associated for some in debates about the public domain for which Wordsworth could be invoked as a poetic spokesman. For a while, this domain also provided another representational locus at the interface between Anglo-American relations: the cultural homeland as national park.

Wordsworth's American visitors, many of whom were writers, sought to represent their own interests in and as American 'cultural and natural' landscapes. Wordsworth and the Lake District provided points of reference and comparison for literature and society, as did Coleridge's concepts of 'clerisy' or 'National Clergy', and national 'Reserve'.[27] Authors and artists helped to authenticate the picturesque, imaginative and touristic value of sites which, in turn, could be useful to democratic government. During his lifetime many notable Americans came to converse with Wordsworth in the Lake District; as he grew older, and in the decade following his death, the area of his inspiration became both a real and an imaginative place of reference and comparison for their own national project. From

---

24 Everett, *Tory View*, p. 204; Coleridge, *Church and State*, pp. 22, 72–3; Williams, *Culture and Society*, pp. 59–62.

25 Williams, *Culture and Society*, p. 64, quoting the 1837 edition of *Church and State*, V.

26 Thompson, *Lake District*, pp. 72–3.

27 Coleridge, *Church and State*, pp. 47, 72.

the 1870s a collection of circumstances brought English and American 'clerisies' together to protect their representational interests, often against new demands for the same sites. Realised 'cultural homelands' developed at this interface.

Americans in Wordsworthshire

Early American tourists to the Lake District visited Wordsworth. The visitors mentioned here were highly-educated, influential men (and occasionally women) who, like Wordsworth, engaged in their country's literary and political nation-building projects. A powerful Boston–Harvard–Concord axis associated with America's political, intellectual and literary elites is discernible. In this *milieu* interests, friendships and family ties frequently overlapped; recommendations, accounts and introductory letters circulated. These religious and educational leaders, writers, critics, artists, politicians and diplomats were neither supplicant and uncritical nor uniform in their views. Through American accounts, a different picture of the Lake District and Britain's stewardship of 'national property' emerged. The Lake District gained a representational role through Wordsworth's poetry and prose for educated Americans. In turn, the area gained further associations from these Americans and their published recollections. Such accounts gave writers a sub-diplomatic role that was sometimes officially recognised.[28]

Several of Wordsworth's American visitors were involved with museum and landscape preservation by the mid-century. The United States' need to represent national, state and federal identities through such cultural institutions gained momentum in the aftermath of its Civil War (1861–65); the fledgling nation had witnessed far greater ravages of the machine and westward expansion. Edward Everett, Unitarian, orator and statesman, helped raise funds to acquire George Washington's home, Mount Vernon ('preserved' in 1851),[29] though there is scant record of his 1818 visit. When minister (ambassador) to Britain (1841–45) he met Wordsworth as Poet Laureate; he became president of Harvard College (1846–49) and Senator for Massachusetts (1853–54). George Ticknor, Harvard's influential European literature professor, followed in 1819, 1835, 1838 and 1849.[30]

The account left by William Ellery Channing, Boston's influential Harvard-educated, Unitarian minister who arrived (1822) with his wife and a letter of introduction, expresses the quasi-religious aura with which Wordsworth imbued the landscape. For Channing the rural parish was sacrosanct. Welcomed by the Unitarian community in Liverpool (where he later ministered, 1857–62), he saw the Lake

---

28   Buzard, *Beaten Track*, pp. 60, 78.

29   P. West, *Domesticating History: the Political Origins of America's House Museums* (Washington: Smithsonian Institution Press, 1999), pp. 1–38; C.B. Hosmer Jr., *The Presence of the Past: a History of the Preservation Movement in the United States before Williamsburg* (New York: G. P. Putnam's Sons, 1965), pp. 29–62.

30   S. Gill, *Wordsworth and the Victorians* (Oxford: Clarendon Press, 1998), p. 14.

District's scenic beauty as a setting for his own heightened emotional experience and a higher order of society. A sacralisation of nature and culture that forms a part of and transcends nationalism found a location at 'Grassmere Water [sic], a sacred spot, a seclusion from all that is turbulent and unholy in life.'[31] Channing's feelings were intensified by 'atmosphere, fogs, and various lights, [that] give to the tops of mountains a visionary, sometimes a mysterious character' where he imagined the human spirit mingled with the divine.[32] By such diffusion of concepts, images and language the potential terrors of 'Sublime' elements were transfused into evocations of a world both material and heavenly, God-ordained and, thus, 'beautiful' and 'natural'. Endorsement of such atmospheric views by a spiritual leader enhanced their value for followers and invoked a sense of modern pilgrimage.

Channing also visited Robert Southey and, in London, met Coleridge who he regarded as one of the greatest influences on his philosophical thought. When young, he had aspired to emulate Coleridge and Southey's unrealised proposal to establish an ideal community in Philadelphia.[33] Similar Utopian visions emerged in Europe; some became associated with historical places and stood for a greater 'whole'. There is no equivalent word in English for 'heimat' or 'fornhem', though examples abound. Channing affirms Wordsworth's poetic invocation of a 'fornhem' (as a sacralised parish) when he describes the area as, 'combining with its natural beauty the most affecting tokens of humanity by its simple cottages and Gothic churches, [that] communicates an inexpressible character of peace and benignity, and of gentle and holy sweetness, to the whole scene.'[34]

Channing's account, later published, depicts Wordsworth as Nature's poetic cleric. He places the poet in the scene, continuing, 'as I descended into Grassmere [sic] near sunset, with the placid lake before me, and Wordsworth talking and reciting poetry with a poet's spirit by my side, I felt that the combination of circumstances was such as my highest hopes could never have anticipated.'[35] The 'poet's spirit' in various invocations could, metaphorically, be taken by successive tourists as they, too, walked around Grasmere with his poetry guiding an emotional, quasi-communal, emotive and scenic journey. Such affect could only be reproduced by

---

31 W. Channing, *William Ellery Channing, D.D.* (Boston: American Unitarian Association, 1880), p. 338; J. Beer, 'William Ellery Channing visits the Lake Poets', *Review of English Studies* 42, 166 (May, 1991), pp. 212–26. L. Levine, *Highbrow Lowbrow: the Emergence of Cultural Hierarchy in America* (Cambridge, Mass: Harvard University Press, 1988), pp. 13–33, 69–81 observes a 'sacralization of culture' around Shakespeare, *c.*1870s. The process had already taken place with 'nature'.

32 Channing, *William Ellery Channing*, p. 341; B.M. Stafford, *Voyage into Substance: Art, Science, Nature, and the Illustrated Travel Account, 1760–1840* (Cambridge, Mass.: MIT Press, 1984), pp. 185–281, 461–70.

33 Beer, 'William Ellery Channing', p. 213; Richard Holmes, *Coleridge: Early Visions* (Harper Collins, 1998), pp. 60-3.

34 Channing, *William Ellery Channing*, p. 342.

35 Ibid.

maintaining the aesthetic (and sensory) construction of the landscape. Maintaining authorial status and 'authored' location were more worldly affairs.

For some, Wordsworth had celebrity status. In 1838 Harvard law professor and abolitionist, Charles Sumner (a powerful senator from 1851) exclaimed (in correspondence that was later published) to George Hillard: 'I have seen Wordsworth! How odd it seemed to knock at a neighbour's door, and inquire, "Where does *Mr.* Wordsworth live?" Think of rapping at Westminster Abbey, and asking for Mr. Shakespeare, or Mr. Milton!'[36] International intellectual property rights were a concern for Wordsworth, and the men discussed copyright issues. Wordsworth fared well in matters of American royalties. From 1802 his poetry, edited by Professor Henry Reed of the University of Pennsylvania, was set by American printers. Reed passed on royalties.[37]

Britain, for Americans, was a place of departure and point of comparison. Early American writers, claiming to be 'more English than their English counterparts' were 'mounting a challenge to the authority of the mother culture' and creating a new, though related, culture more appropriate to their ideals, history and place.[38] Ensuing visits became less laudatory as Americans engaged in nation-building and relations with Britain became strained. William Cullen Bryant, the so-called American Wordsworth, left scant record of his 1845 visit, perhaps indicative of dwindling popular interest in the ageing English poet.[39] Regarded in England as 'America's only significant poet, its leading journalist, and its distinguished civic leader,' Bryant already knew Harriet Martineau, the social observer and Ambleside resident to whose *Society in America* he largely owed this reputation.[40] Martineau, a utilitarian advocate of industrial modernity, was in some senses the antithesis of Wordsworth. As America's cultural hub moved from Boston to New York literary critic Margaret Fuller, who visited Wordsworth in 1846, went with it. She considered Wordsworth an adaptable role-model for modern American poets but not his view of nature. 'The "beauties of nature" never could console me for any ill,' she wrote in 1830, dismissing as imbecilic the 'horror … of the startling and paradoxical' that she claimed 'admirers of the great poet whom I have known in these parts' derived from Wordsworth's natural theology.[41]

Issues of Englishness and its relationship to England became important to the former colonies in developing independent identities. Writers such as Emerson, Washington Irving and Nathaniel Hawthorne (among others) represented

---

36   Sumner to Hillard, 8 September 1838. *Memoir and Letters of Charles Sumner*, ed. E. L. Pierce, 4 vols (London, 1879–93), vol. 1, pp. 355–6, quoted in Gill, *Wordsworth*, p. 14.

37   Gill, *Wordsworth*, pp. 9, 382.

38   Tennenhouse, *Importance of Feeling English*, pp. 17–18.

39   G.H. Muller, *William Cullen Bryant, Author of America* (Albany: State University of New York Press, 2008), pp. 183–6; W.C. Bryant II, *The Letters of William Cullen Bryant, Vol. 2, 1836–1849* (New York: Fordham University Press, 1977), p. 342.

40   Bryant, *Letters*, p. 342.

41   Capper, *Margaret Fuller*, p. 91.

alternative national identities through language and place.[42] Place, expressed in historical, geographical and quasi-religious terms, and language-use, gained significance in authenticating claims to distinction and hierarchy. Influential essayist Ralph Waldo Emerson, the Harvard-educated, Transcendentalist and Concord resident, published his 1833 and 1848 visits in *English Traits* (1856). The first focused on Coleridge's and Wordsworth's interest in America.[43] He returned to Rydal Mount with Harriet Martineau as a more comparative critic of English society and more engaged in American romantic nationalism.[44] By then Wordsworth was an old man and Emerson compared what he saw as England's rigid, hereditary, even feudal society with the wider possibilities available to Americans. The idea of a hereditary community as envisaged among 'the small freeholders in Westmoreland' by Wordsworth as socially stabilising was translated by Emerson as picturesque quaintness and resistance to change.[45]

Several authors held diplomatic posts; Nathaniel Hawthorne toured the area in July 1855 while consul in Liverpool. Hawthorne who, like Longfellow, graduated from Bowdoin College in Maine, rented (1842–44) the Emerson family's former manse in Concord, a rural colonial town near Boston and erstwhile site of revolutionary fighting then gaining recognition as a literary colony.[46] Hawthorne memorialised the site of his writing and the Emerson family home in *Mosses from an Old Manse* (1846). In Emerson, America's modern 'clerisy' had a direct, if recent, ancestry with which to legitimate claims to cultural ministry that was also tied to nation-building events.[47] With Emerson's *Nature* (1836) and David Henry Thoreau's *Walden, or Life in the Woods* (1854), Concord's sense of place gained a peaceful, rustic-picturesque mantle that was disturbed by the railway that passed by Concord's wooded outskirts from 1844. Preservation began privately that year as Emerson began buying and reserving adjacent land.[48] Thoreau recorded his Transcendentalist experiment in a small cabin near Walden Pond on Emerson's land. Wordsworth, viewed by some in the States as a 'hermit among the mountains', provided one role-model.[49] Thoreau believed, 'To live a good old age such as the ancients reached, serene and contented, dignifying the life of

---

42 Tennenhouse, *Importance of Feeling English*, pp. 17–18.

43 Emerson, *English Traits*, p. 10.

44 Ibid., pp. 165–6.

45 Ibid., p. 62.

46 The house belonged to Emerson's uncle, Revd Samuel Ripley.

47 The Old Manse was 'witness' to Revolutionary events, as was Emerson's grandfather, Revd William Emerson, who built the private parsonage. See also Turner, *Liberal Education*, p. 12.

48 I am grateful to Brian Donahue for this information. Emerson continued to buy land as the railway company developed an amusement park in 1866, thus protecting the site of the two men's 'nature' inspiration.

49 *Southern Literary Messenger* (1837), quoted by M. Reed, 'Contacts with America', in N. Clutterbuck (ed.), *William Wordsworth 1770–1970: essays of general interest on Wordsworth and his time* (Grasmere: Trustees of Dove Cottage, 1970), p. 35.

man, leading a simple, epic country life in these days of confusion and turmoil, – that is what Wordsworth has done ... Heroism, heroism is his word, his thing.'[50] *Walden* simultaneously invoked Concord's historical associations and a modern Edenic-Arcadia while defining place through closely observed 'nature'. *Walden* bolstered Thoreau's status as cultural (if hermitish) cleric and further authenticated Concord's growing identity as a 'natural' and cultural homeland for New England and the 'new' nation, as had Wordsworth with the Lake District. However, a new attitude to 'the machine in the garden' was also evident in Thoreau's writing; the railway on Concord's periphery was both a sign of unity as well as symbolic of a mechanistic intrusion in pastoral life.[51]

Hawthorne's tour, written during 1853–58 and published posthumously as *The English Notebooks* in 1870, expresses further differences as he compares place and society while adding temporality.[52] Five years after Wordsworth's death, Hawthorne visited Grasmere and Rydal Mount where he and his wife sought a sprig of ivy (for remembrance) from the poet's garden wall. A new particularity of place accompanies a new sense of authorial individuality; Hawthorne's own adventure predominates. Where Sumner had expressed awe, humour defused wonder; while concepts of place, displacement and misplacement, plus American pluck and luck, are woven into a jovial, comparative narrative. Neither vicar, nor poet but a gardener informs them that their target was Wordsworth's neighbour's house, a mistake that enables a guided tour of 'Wordsworth's' garden. Hawthorne compared the wider landscape in terms of prospects and resource management. Enjoying picturesque views and textured signs of lives past, he disparaged Rydal Water's size finding it, 'not a permanent body of water, but a rather extensive accumulation of recent rain.'[53] His main criticism concerned land-ownership and being charged to see Aira Force; echoing Wordsworth, he deemed 'the Beautiful [to be] the property of him who can gather it and enjoy it.' Considering 'it ... very unsatisfactory to think of a cataract under lock-and-key', he reflects growing interest in waterfalls focusing around Niagara Falls where both British and American responses to stewardship of natural (and cultural) resources were on display (discussed below).[54]

Wordsworth, together with his fellow Lake poets and artists like Turner, 'authenticated' an appearance for the Lake District in association with specific values

---

50 D.H. Thoreau, *The Journal of Henry D. Thoreau in fourteen volumes bound as two, Vols I–VII, (1837–October 1855)* (New York: Dover, 1962), p. 120 (1845–47, I: 408).

51 Walden Pond's size, relative to other bodies of water on the Continent, merited its definition as a pond; in English terms it would be a lake. Marx, *Machine in the Garden*, pp. 220–21, 242–46.

52 N. Hawthorne, *The English Notebooks, 1853–1856*, ed. T. Woodson and B. Ellis (Columbus: Ohio State University Press, 1997), pp. 231–80.

53 Ibid., p. 247.

54 Ibid., p. 260.

and, often, with powerful emotions and half-expressed memories of religion.[55] After his death, Wordsworth's 'sacred' association with the place was reinforced by his burial in St Oswald's Churchyard, Grasmere. British and American contributions recognised the connection through memorials: a monument in Grasmere parish church and a window in the new St. Mary's Church, Ambleside (1854). American contributions of 'about seven hundred dollars' harnessed by Henry Reed were represented as paying – separately – for the window.[56] Connotations of Anglo-American harmony, reinforced through religious fellowships, were important for post-Civil War relations. Subscribers included Emerson and Henry Wadsworth Longfellow, Harvard professor (1836–54) and popular poet, thereby representing an American literary establishment's contribution to the memorial of a canonical figure of modern English literature.[57] Americans were creating their own identities in and of the Lake District; James Russell Lowell, Harvard professor of Belles-Lettres, poet, essayist and diplomat who edited American editions of Wordsworth's works, named the area 'Wordsworthshire' in his 1854 'Sketch of the poet's life'.[58] Lowell, another of the Cambridge 'enclesia', studied law at Harvard with Channing and would become minster to Britain (1880–85), (when he returns to this narrative).

From Cultural Homeland to National Park

In the early-nineteenth century natural theology validated the Lake District's scenery. By mid-century this vision was less sustainable. Ruskin, for whom Nature was the transcendent artist, complained to Henry Acland, 'If only the geologists would leave me alone, I could do very well, but those dreadful hammers! I hear the clink of them at the end of every cadence of the Bible verses.'[59] The Lake District was more than an imaginary space. It was a place where the values and

---

55   For a discussion of the authenticity effect see Buzard, *Beaten Track*, pp. 172–92.

56   *Transactions of the Wordsworth Society* 2 (1882), p. 88; B. Falbo, 'Henry Reed and William Wordsworth, an editor-author relationship and the production of British Romantic discourse', *Romantic Textualities, Literature and Print Culture, 1780–1840* 15 (Winter, 2005): pp. 28, 46 at note 1; Gill, *Wordsworth*, pp. 36–7. Memorial windows became associated with American contributions; for other examples see M. Hall and E. Goldstein, 'Writers, the Clergy, and the 'Diplomatization' of Culture: Sub-Structures of Anglo-American Diplomacy, 1820–1914', in J. Fisher and A. Best (eds), *On the Fringes of Diplomacy: Influences on British Foreign Policy, 1800–1945* (Farnham: Ashgate, 2011), pp. 127–54.

57   Karbiener, 'Intimations of Imitation', in Pace and Scott, *Wordsworth*, p. 150.

58   Wordsworth, *The Poetical Works*, unsigned 'Sketch of Wordsworth's life' by J.R. Lowell, vol. 1, p. xxxviii. I am grateful to Polly Atkin and Jeff Cowton for pointing this reference out to me.

59   Ruskin to Henry Acland, 24 May 1851, quoted in J.M.I. Klavier, *Geology and Religious Sentiment: the effect of geological discoveries on English society and literature between 1829 and 1859* (Leiden, New York, Köln: Brill, 1997), p. xi. I am grateful to Bernard Richards for this reference. Ruskin's influence lies beyond the scope of this chapter.

visions of authorship and a cultural clerisy could be affirmed by polite tourism. Its 'authenticity' as Wordsworth knew it was increasingly subject to change.

Author's tours were a growing phenomenon in Anglo-American relations.[60] The popular recognition of Longfellow's distinctive American literature helped cultural relations. Longfellow enjoyed celebrity status. His much-publicised (1868–69) 'tour' gave the Lake District further sub-political status and representational value for authors and for modern English literature's distinctiveness (Figures 5.1 – 5.3). Reported in *The Times*, Longfellow's arrival in Britain featured in the same *New York Times* news item as the new American minister's arrival, and received more coverage.[61] Longfellow's tour had prestige; he was received by the Queen and given honours by Oxford and Cambridge Universities; his popularity ensured huge receptions at railway stations. The tour encompassed significant literary sites: the Lake District, Abbotsford and Stratford-upon-Avon.[62] Celebrity visits, recorded by their autographs, added further interest for visitor-readers.

Literary sites accrued utility for representational, educational purposes as Wordsworth's poetry was included on American school syllabi.[63] Industrialisation brought a rise in popular entertainments. Sometimes 'highbrow' and 'lowbrow' cultures could co-exist but, increasingly, cultural distinction was marked by divisions of place as well as content.[64] Young Alice Longfellow left a rare account of 'show-business' entertainment in Wordsworthshire. After viewing Wordsworth's church pew, Alice enthused about 'a most comical show in a cart, where we saw animals; a pony; and an African dance a war dance, and eat raw liver'.[65] Such popular 'African shows' were more usually associated with urban settings. They presented a more global vision of the world's population showcased as unusual or 'exotic' rather than as 'folkish'.[66] These were among the new aesthetics of the expanding industrial world rather than of local agrarian communities that some sought to set aside for rustic picturesque viewing.

---

60   Hall and Goldstein, 'Writers'.

61   *Times*, 10 June 1868, p. 5, col. C; *New York Times*, 7 July 1868, p. 2

62   Sir Walter Scott's library had been saved intact. Shakespeare's Birthplace had already been preserved in situ and against American show-business interest, see M. Hall, 'Plunder or Preservation? Contesting the Anglo-American Heritage in the Later Nineteenth Century', in P. Mandler and A. Swenson (eds), *From Plunder to Preservation: Britain and the Heritage of Empire*, Proceedings of the British Academy (Oxford: Oxford University Press, forthcoming).

63   E. Fay, 'Wordsworth, Bostonian Chivalry, and the Uses of Art', in Pace and Scott (eds), *Wordsworth*, pp. 177, 194 at note 2.

64   Levine, *Highbrow*, pp. 56–69, 76–8.

65   'First Journey to Europe, 1868', 9 June 1868. Alice Mary Longfellow Papers, National Park Service, Longfellow National Historic Site, Cambridge, MA, Longfellow House Trust Records, Box 19, Folder 1.

66   For Africans in literature and African shows see H.L. Malchow, *Gothic Images of Race in Nineteenth-Century Britain* (Stanford: Stanford University Press, 1996); G.H. Gerzina, *Black Victorians/Black Victoriana* (New Brunswick, NJ: Rutgers University Press, 2003).

Figure 5.1   Photograph of Grasmere from the Wishing Gate, taken during Henry Wadsworth Longfellow's tour

Source: Longfellow Family Photograph Collection, 1845–1972, Album 10 – English Scenes Album. © Longfellow House – Washington's Headquarters National Historic Site, courtesy of the National Park Service.

Figure 5.2   Photograph of Rydal Water from Wordsworth's Station, taken during Henry Wadsworth Longfellow's tour

Source: Longfellow Family Photograph Collection, 1845–1972, Album 10 – English Scenes Album © Longfellow House – Washington's Headquarters National Historic Site, courtesy of the National Park Service.

Figure 5.3   Photograph of Ullswater, taken during Henry Wadsworth Longfellow's tour

Source: Longfellow Family Photograph Collection, 1845–1972, Album 10 – English Scenes Album © Longfellow House – Washington's Headquarters National Historic Site, courtesy of the National Park Service.

Sites of Literary Friendship

Lake District preservation gained momentum with the formation of the Wordsworth Society. In 1883 Ruskin suggested that its 'grand function' was 'to preserve as far as possible in England the conditions of rural life which made Wordsworth himself possible', looking beyond the visual picturesque to a social and environmental engagement that included imaginative stimuli for literary production.[67] The Society aspired to protect and direct Wordsworth's reputation by bringing out editions of his works, acquiring Dove Cottage and protecting the landscape as Wordsworth knew it.

---

67   John Ruskin to William Knight, 3 April 1883. Pierpont Morgan Library. Printed in part in *Transactions of the Wordsworth Society* 5 (1883), p. 4, quoted in Gill, *Wordsworth*, p. 246.

During the 1880s and 1890s protecting literary sites, representing Anglo-American friendship and issues of imperial federation co-aligned.[68] Among the Wordsworth Society's founder members were James, later Viscount Bryce, Canon Hardwicke Rawnsley and Revd Stopford Brooke: all were active authors with American readers; all had personal connections to the Boston–Cambridge area of Massachusetts. Bryce is a complex figure; a preservationist active in the Lake District and Scotland, he was an Oxford law don, Liberal M.P., spokesman for Gladstone's concept of 'English-speaking peoples' and prolific author.[69] While preparing his influential *American Commonwealth* (1888) he visited Harvard and toured literary sites with Longfellow.[70] Seeing the United States as a successful continuum of Anglo-Saxon government traditions, Bryce re-orientated Britain's popular perception away from the American War of Independence to a common, colonial past and provided an intellectual and cultural framework for Anglo-American legal and political friendship. He became ambassador to the United States, 1907–13.

Towards a Lake District 'National Park'

The campaign to protect the Lake District's landscape began in earnest at the Wordsworth Society's fourth meeting. Following Stopford Brooke's paper on Wordsworth's *Guide*, Rawnsley proposed the formation of a Permanent Lake District Defence Society.[71] The Thirlmere reservoir works were an important indication of the need for a group that could help to mediate, though not necessarily to halt, change. There were other factors. The successful institution of nature-parks in the United States provided important representational precedents; Rawnsley noted park projects in Yosemite (reserved in 1868); in the Adirondacks, where campaigns for sustainable development in conjunction with scenic reservation had begun; and Yellowstone reserved for its natural wonders by the federal government in 1872.[72] Additionally, the campaign to reserve Niagara's banks,

---

68   Hall and Goldstein, 'Writers'; M. Hall, 'The Politics of Collecting: The Early Aspirations of the National Trust, 1883–1913', *Transactions of the Royal Historical Society*, 6th ser., 13 (2003), pp. 345–57.

69   F.H. Herrick, 'Gladstone and the concept of the "English-Speaking Peoples"', *Journal of British Studies*, 12, 1 (1972), pp. 150–6; H. Tulloch, *James Bryce's American Commonwealth: the Anglo-American background* (Woodbridge: Boydell Press, 1988).

70   Bryce visited Longfellow and the Paul Revere House, 1870. Bryce to Henry Wadsworth Longfellow, 28 August 1870, Longfellow Papers, bMS AM 1340.2 (819), Houghton Library, Harvard University, Cambridge, Mass.

71   *Transactions of the Wordsworth Society* 5 (1883), pp. 45–60; Gill, *Wordsworth*, pp. 257–9;

72   For the Adirondacks see S. Hays, *Conservation and the Gospel of Efficiency: the progressive conservation movement 1890–1920* (Cambridge, Mass: Harvard University Press, 1969), pp. 15, 189–92.

islands and waterfalls provided an important model and common cause. In all of these American sites the idea was not to prevent change but, to ameliorate it.

Niagara Falls, an icon of the continent made famous by Frederic Church's painting *Niagara* (1857), straddles the border between the United States and Canada, for which Britain spoke in matters of foreign policy. It was subject to increasing industrial demands including from show-business entrepreneurs.[73] Lord Dufferin, as governor-general of Canada, expressed the desire to reserve this area as 'an international park' (1878); to do so required on-going Anglo-American cooperation. Church prompted a campaign to protect the view; it was led by New England's Cambridge 'clerisy', Harvard's art history professor and Ruskin's friend, Charles Eliot Norton (who, family legend said, was bounced on Wordsworth's knee as a baby) and the landscapist Frederick Law Olmsted. 'Niagara reservationists', as they were known (perhaps again echoing Coleridge), gained powerful British support, including from Ruskin and Thomas Carlyle, with political backing behind the scenes.[74] As picturesque viewing parks were completed, in 1885 on the American side and 1888 on the Canadian, the Niagara reservation campaign coincided with early campaigns for rustic picturesque parks in the Lake District.

A direct link between New England's 'enclesia', American government and the Wordsworth Society was forged in 1884 when James Russell Lowell, America's minister to Britain, was elected its president.[75] It is inconceivable that he accepted this representational role without his government's approval. Though only his inaugural lecture was reported, this was an expression of common cultural interest. Although Wordsworth's lens was less useful to America's western nature reserves, his approach to nature still echoed in calls to protect scenery along the eastern seaboard, notably at Mt Desert Island, Maine with which the Cambridge reservationists had close ties.[76] With Lowell's acceptance of the Wordsworth Society presidency, as with the Wordsworth window, memorialisation of an association of authorship with place could also represent friendship and a form of cultural federation.

The next significant visitor to Wordsworthshire was Olmsted's young associate Charles Eliot, son of Harvard president, Charles W. Eliot, and Charles Eliot Norton's cousin. As Eliot made his 'giro of Europe' (1885–86), he visited and corresponded regularly with family friend James Bryce. On 22 July 1886 after breakfast with Bryce, Eliot picked up 'reports' from the Commons Preservation Society (C.P.S., founded 1865). These included (Sir) Robert Hunter's 1884

73   See M. Hall, 'Niagara Falls: Preservation and the Spectacle of Anglo-American Accord', in Hall, *Towards World Heritage*, pp. 23–44.

74   Turner, *Liberal Education*, p. 29.

75   *Transactions of the Wordsworth Society* 5 (1883), pp. 12–24.

76   D. Haney, 'The Legacy of the Picturesque at Mount Desert Island', *Journal of Garden History* (October–December 1966), pp. 274–93. I am grateful to David Haney for this information.

proposal for a land-holding preservation organisation that Bryce had had printed for circulation around the C.P.S. and which subsequently was to result in the 'The National Trust for Places of Historic Interest and Natural Beauty'. Eliot's copy of Hunter's report remains in his library.[77] It undoubtedly provided a model for Eliot's Massachusetts-based Trustees of [Public] Reservations (founded 1890) which, in turn, gave legal precedence (through their shared and cross-referential use of case law) for the National Trust. Bryce, as well as Hunter, Rawnsley and Octavia Hill, was among the Trust's founders. These two voluntarist organisations had close ties to the state and could wield some political influence; the National Trust's founding Council had a distinct Liberal character with around 45 Liberal M.P.s or members of the House of Lords.[78] Both aspired to protect mytho-poetic cultural homelands as landscape parks, together with authors' houses and buildings with other historical associations that added meaning.[79]

Eliot was a Wordsworthian. Attending Revd Stopford Brooke's sermons he recorded, 'Nature ... to those who come to her and say – I love you – gives joy and delight unspeakable. Those who know Wordsworth know what this is.' In August, he enjoyed a 'fabulous view of ... Windermere' and travelled by coach from 'Ambleside – and so past Rydal Water – on past Grasmere ... past Thirlmere, and works just entered on for taking water hence to Manchester.' Walking to Scale Force he found 'Scenery of a very distinct type – and of its sort the perfectest imaginable.'[80] The use of water resources, including the reservation of waterfalls in landscape parks was topical; the optimal way to protect scenic interests was to institute parks. At the C.P.S. secretary's suggestion Eliot visited Rawnsley who, with Bryce, was campaigning to prevent railways from intruding into scenery and reserve access to footpaths and common land in the Lake District.[81] It is worth quoting Eliot's Journal entry for 7 August at length; he notes that Rawnsley was:

> Sec[retary] and prime mover in Lakeland Defence Association – wh[ich] has fought off two railway schemes and done other service. Introduced myself as an American much interested – got him talking – closing of ancient footways chief trouble at present – done right and left by nouveau riche new proprietors – and hard to prevent because costly – burden of proof strangely lying with

---

77   Charles Eliot's Journal in Europe 1885–86, 22 July 1886, Alexander Yale Goriansky Papers (hereafter Eliot's Journal). A copy of Hunter's published lecture is among the Charles Eliot Papers, CE NAB 6010, Loeb Design Library, Harvard University, Cambridge, Mass.

78   Hall, 'Politics of Collecting', pp. 345–57.

79   See M. Hall, 'Affirming Community Life: Preservation, National Identity and the State, 1900', in C. Miele (ed.), *From William Morris: Building Conservation and the Arts and Crafts Cult of Authenticity 1877–1939*, Studies in British Art 14 (New Haven and London: Yale University Press, 2005), pp. 129–58.

80   Eliot's Journal, 4 and 5 August, 1886.

81   Murphy, *Founders*, pp. 85–8; Taylor, *Claim on the Countryside*, p. 133.

'the public'. ... New law wanted to enable local authorities to fight the battles, instead of Rawnsley and Co. Parliament will soon take this up – along with Mr B[ryce's]'s Scottish Mnts[mountains] Bill. In other matter Parl[iament] has already affirmed principle of real value of scenery – has refused to charter railways wh[ich] would have injured scenery – and has required Manchester water works people to save soil wherewith to recover their masses of tunnel debris etc. Mr. R[awnsley] asked after my work in America – if I were in the Law or what? – and hearing my trade made me sit down again and tell him about *Yellowstone, Niagara* etc. – then looked at his old vicarage garden – and fled for noon train.[82]

Their conversations about Yellowstone and Niagara are significant; by the 1880s, both of these resources were used to represent cultural federation within the United States, though Niagara's inclusion in an 'emerald necklace' of parks across the continent could also perhaps suggest Canada's inclusion. Landscape parks were important ways of expressing civic, state and federal identities in the New World. Yellowstone was made a federal reserve in 1872 yet, although it was colloquially known as a 'national park' among other descriptions, Sydney, Australia was the first to designate 'The National Park' in 1879.[83] The first 'official' National Park in the United States was Mesa Verde (1906). Public, landscape parks grew popular in the white settler dominions, suggesting a form of cultural federation and democratic representation that called into question the mother-country's cultural leadership, set up a cultural dichotomy that distinguished the 'new' world from the 'old' and seemingly separated 'nature' from 'culture'.

Like Yellowstone and, indeed, the Niagara area, the Lake District straddles more than one bureaucratic district; however, even more than at these places, there were existing property rights and business interests. In 1887 a Petition against the Ambleside Railway Bill, prompted by Rawnsley and supported by Bryce, claimed that 'the Lake District at present serves the purposes of a great National Park, and is annually visited (on account of its exceptional beauty) not only by thousands of people from all parts of the United Kingdom, but by great numbers from the British Colonies, the United States of America, and the Continent of Europe'.[84] The issue was championed by the *Lancaster Observer*; describing the Lake District as a 'National Treasure' like Yellowstone, it prompted Ruskin's concurrence 'in the recommendation that the whole Lake District should be bought by the nation for

---

82   Eliot's Journal, 6 and 7 August, 1886 (my italics).

83   *New South Wales Government Gazette* 148 (26 April 1879), pp. 1923–4. I am grateful to Richard White for this reference.

84   Papers of the Lake District Defence Society, Cumbria Record Office, Carlisle. DSO/24/3. Petition addressed to the House of Lords, signed predominantly by Cambridge University faculty members.

itself'.[85] If successful the 'old world', 'new world' park dichotomy could become a dialectic integrating poetic echoes of Wordsworth and the other Lake poets, including Coleridge.

The railway scheme was dropped but a campaign had begun to protect sites associated with Wordsworth's life and work. Its first success was the Wordsworth Society's acquisition of Dove Cottage (1900), tying authorship to place. The railway could do more than transport people to a place; it could transport a place to another people. Acquiring English authors' homes gained momentum during the 1880s when an 'American showman' targeted Milton's Cottage, Chalfont St Giles for purchase.[86] As museums, these houses had broad representational utility and, while pleasure-leisure-zones assist in levelling social distinction, Coleridge's 'National Clergy' (sometimes with aristocratic philanthropic assistance) preferred to utilise such sites for civic and educational purposes. Dove Cottage, according to 'Pictures of Travel' by 'an American' in *The Ladies' Repository* (August 1862), was 'a little story-and-a-half cottage, mortared and whitewashed, looking no better, except for whitewash, than many of the rude huts around it. A little room used as his study is now a little huckster's shop.'[87] To the 'clerisy', such commercial use was inappropriate. Revd Brooke recorded his aspiration to acquire the Cottage 'for the eternal possession of those who love English poetry all over the world' and less expansively 'for the pleasure and good of the English race'.[88] Some contributions came from the United States. *The Nation* (that Charles Eliot Norton helped found in 1865) in the United States reported Professor Knight's donation of Wordsworthiana to 'the national museum, as [Dove Cottage] must now be called'.[89] The journal also noted the museum's popularity among Americans and those from other English-speaking countries.

Dove Cottage provided a locus for 'reserving' the Lake District as a Wordsworthian 'national property'. Campaigns continued through the National Trust. The organisation maintained links with the United States and Canada during its early years; Rawnsley toured the East Coast and Niagara Falls in 1899; a short-lived branch of the Trust was established in Concord and Washington in 1900–1904.[90] The fledgling Trust had clear associations with noted Imperial

---

85   J. Ruskin, *The Works of John Ruskin*, ed. E.T. Cook and A. Wedderburn, 39 vols (London, 1903–12), vol. 34, p. 604, quoted in Gill, *Wordsworth*, p. 259.

86   *Daily News* (3 May 1887), p. 3; *New York Times* (15 May 1887).

87   I am grateful to Jeff Cowton (Dove Cottage) for this reference.

88   S.A. Brooke, *Dove Cottage, Wordsworth's Home from 1800–1808* (London and New York: Macmillan and Co., 1890), p. 14; Gill, *Wordsworth*, pp. 245–6.

89   *The Nation* 76, 1725 (21 July, 1891), p. 52; Turner, *Liberal Education*, p. 197.

90   E. Rawnsley, *Canon Rawnsley, An Account of his Life* (Glasgow: Maclehose, Jackson and Co., 1923), p. 258. See also J.K. Walton, 'The National Trust: preservation or provision?' in M. Wheeler (ed.), *Ruskin and Environment* (Manchester: Manchester University Press, 1995), pp. 144–64.

Federationists including Canada's park-making governors-general.[91] Lord Dufferin was its second president; his successor in the Trust was Princess Louise whose husband, Lord Lorne, was Dufferin's successor in Canada suggesting English-speaking cultural federation was topical. Protecting cultural landscapes as national parks and nature reserves represented a federal connection as 'natural'. It is no coincidence that adjacent to Canada's first, Banff Hot Springs Reserve (1885, now Banff National Park) is the town and lake of Windermere.

Seen retrospectively, representing Wordsworthian reserves as national parks was, perhaps, a step towards instituting a Lake District National Park. Brandlehow, acquired by the Trust in 1902, was described as England's 'national park' by Lord Lorne, speaking on the Princess's behalf, at the opening. He discussed their work instituting similar parks in Canada following America's precedent. Lord Lorne highlighted the Lake District's association with scenery and literature (also literature's scenery). Stressing the need to set land aside in Britain 'for national purposes and public recreation', he declared,

> The object of the [National] Trust ...[is] not only to preserve places of beauty and historic interest for the nation, but also to preserve this country of ours for the delectation of future generations. ... [England] was the mother of those nations which might in future rank with her, and it [is] of the utmost importance that the charms of the country – the places of beauty and historic interest which we possess – should be preserved, not only for the people of this country, but for the inhabitants of the colonies, the great kindred nation on the other side of the Atlantic who in the future might resort here as the home of their leisure and as the place where they would enjoy what they had acquired by their labours.[92]

Two years later the appeal for the Gowbarrow Estate and Aira Force went further by suggesting it would form 'the finest national park in England'.[93] Nevertheless, this was still a sort of 'national park' within the Lake District and not a Lake District National Park. As Dorothy Wordsworth recorded, it was at Gowbarrow where she and her brother saw the flowers that inspired his famous poem, 'Daffodils'. Thus, the relationship of authorship and site was expanded beyond the home to include places of inspiration and activity. Rawnsley invited Woodrow Wilson, then a Princeton professor visiting the site of his grandfather's ministry, to its opening. Interestingly as President, Wilson approved legislation that created

---

91   J. Mackenzie, *Propaganda and Empire: the manipulation of British public opinion, 1880–1960* (Manchester: Manchester University Press, 1984), pp. 151–2 notes important members of the Imperial Federation League (1884–93); these included Sir John Lubbock and the Duke of Devonshire who were influential in the National Trust, as was the Duke of Argyll.

92   *The English Lakes Visitor and Keswick Guardian* (18 October 1902), p. 5.

93   Quoted in Rawnsley, *Canon Rawnsley*, p. 112.

America's National Park Service (1916), thus aligning popular perceptions with formal representation.

At their best, landscape reservations provide places and spaces where the past can be renegotiated with a future in mind, change can be mediated, and the idea of common good can be imagined. In the two Lake District 'national parks', Wordsworth's inspiration and authorship was represented, respected and reserved as 'cultural homelands' for an English-speaking, reading, viewing and touring public at home and abroad. As this chapter has argued, American engagement with Wordsworthshire and, indeed, Britain's engagement with North America contributed to this goal. They provided mytho-poetic, aestheticised middle grounds not only between the imagination and reality of the reader and 'nature' but, between the 'mother country' and its white-settler, English-speaking communities. Further, they suggested a federated form of natural and cultural 'national properties'. Here, site affirmed authorship, and authorship authenticated site. Such gestures affirmed English-speaking cultural diplomacy, Wordsworth's writings as a guide to appreciation of picturesque nature and, perhaps, a limited recognition that nature's bounty has limitations.

PART II
Lake District Tourism Themes

Chapter 6

The Imaginative Visitor: Wordsworth and the Romantic Construction of Literary Tourism in the Lake District

Keith Hanley

'The Region of Wordsworth's Song'

The Lake District became known as "Wordsworthshire"[1] because, as Edward Thomas wrote, the link between writer and a specific place is nowhere stronger than in this case:

> It is more natural and legitimate to associate Wordsworth with certain parts of England than any other great writer. And for three reasons: he spent the greater portion of his life in one district; he drew much of his scenery and human character from the district and used its place-names freely in his poems; and both he and his sister left considerable records of his times and places of composition.[2]

The singularity of his attachment then and since derives not only from the place, but also from his historical moment. As W.G. Collingwood wrote, 'If Wordsworth had been born a century later he could not have been Wordsworth'.[3] Collingwood was referring to the visible balance between nature and humanity, which had changed markedly by Collingwood's own day. Along with those changes had come radically different social and political assumptions from those which informed the poet's revolutionary lifetime. Place and time were locked together for Wordsworth in the way that his 'spots of time' in *The Prelude* exactly pin-pointed the co-ordinates of his most intense imaginative experiences.

Wordsworth's historical geography resulted from his having been born and bred in the English Lake District, and having thereafter been traumatised by a tumultuous revolutionary period, in France and industrial England, which threatened to undermine everything for which that local culture stood. In his

---

1 See Melanie Hall's chapter, above; E. Robertson, *Wordsworthshire. An Introduction to a Poet's Country* (London: Chatto & Windus, 1911).

2 E. Thomas, *A Literary Pilgrim in England* (Oxford: Oxford University Press, 1980), p. 261.

3 W.G. Collingwood, *Lake District History* (Kendal: Titus Wilson, 1925), p. 1.

writings, his early experiences of the Lakes became the foundation for spiritual discourses and a conservative anti-capitalist ideology. Representations emerged from the pre-Romantic cult of the Picturesque which had specialised in the interplay between reality and artistic manipulation according to the rules of its chief proponent, William Gilpin.[4] The watercolour medium, for example, perfectly matched the kind of effect which Wordsworth himself described in his *Guide to the Lake District*: 'an elevation of 3,000 feet is sufficient to call forth in a most impressive degree the creative, and magnifying, and softening powers of the atmosphere'.[5] He was intimately familiar with the conventional agenda of picturesque tourism, which encouraged the organisation of natural experience for pleasure and entertainment, so that artifice and nature were happily fused. Visitors, for example, delighted in extending natural echoes by firing their fowling pieces over the lakes, and in the late eighteenth century the natural echoes of Ullswater and Derwentwater were enhanced by the use of cannon fire.

By 1799, however, when Coleridge accompanied Wordsworth on his first visit to the Lakes, the practices of what the former called 'Pikteresk Toorists' had become formulaic, and Wordsworth was keen to effect a deeper encounter, conducting his friend off the beaten tourist routes, 'following tracks and ancient roads ... through the very heart of the district, known to none but the local people'.[6] He was eager to provide insights into the life of local communities which were authentic, in ways previously unobserved, so that the inhabitants were unaware of being objects of detached examination. In his Preface to *Lyrical Ballads* the following year Wordsworth described the kind of local culture of '[l]ow and rustic life' which his poetry aimed to convey to a general readership:

> because in that situation the essential passions of the heart find a better soil in which they can attain their maturity, are less under restraint, and speak a plainer and more emphatic language; because in that situation our elementary feelings exist in a state of greater simplicity and consequently may be more accurately contemplated and more forcibly communicated; because the manners of rural life germinate from those elementary feelings; and from the necessary character of rural occupations are more easily comprehended; and are more durable; and lastly, because in that situation the passions of men are incorporated with the beautiful and permanent forms of nature.[7]

---

4   K. Hanley, *John Ruskin's Romantic Tours 1837–38: Travelling North* (Lampeter: Edwin Mellon, 2007), pp. 39–40.

5   W. Wordsworth, *William Wordsworth's Guide to the Lakes, The Fifth Edition (1835)*, ed. E. de Sélincourt (Oxford: Oxford University Press), p. 102.

6   M. Lefebure, *The Illustrated Lake Poets: Their Lives, Their Poetry and the Landscape that Inspired Them* (London: Tiger Book International, 1992), pp. 12, 15.

7   R.L. Brett and A.R. Jones (eds), *Lyrical Ballads* (London: Routledge, 1968), p. 245.

But by the time Wordsworth was writing his ballads, he himself was a returning visitor to the scenes of his past and had come to appreciate that they were different in that they had by then acquired a consciously representational significance. In 'Home at Grasmere', he celebrated the exemplary success of re-inserting himself into his rural culture of origin, but the intellectual superstructure he now attributed to the district depended on a narrative of separation and return, whereby he both belonged to that community and also could share an external gaze on it. As he had described in 'Lines Written a few miles above Tintern Abbey', in reference to his former visit there five years previously, his youthful passion for picturesque sensationalism had matured into meditative values and 'something far more deeply interfused'.[8]

In effect, Wordsworth had come to scrutinise his Lakes past for values which could repair the trauma of both his political disenchantment with the French Revolution and his apprehension of the social and economic effects of the industrial revolution.[9] As he was reconstructing his original community to confront terrorism and tyranny on the Continent, that way of life was becoming increasingly under threat of disintegration from within his own nation, so that he was challenged also to connect the experience of local attachment with a Burkean counter-revolutionary discourse. The precariousness of his alternative tradition is treated in the tragic pastorals of the second edition of *Lyrical Ballads* (1800), intended as part of a series of poems about real contemporary shepherds. They narrate, for example, the total extinction of a hereditary way of life in 'The Last of the Flock', the lure and irreconcilability of modern commerce in the fate of the old shepherd's son, Luke, in 'Michael', and the rupture of community into alienation in the surviving sibling, Leonard Ewbank, in 'The Brothers'.

The last poem tells the story of Leonard's homecoming after twenty years' absence to resettle in his village only to learn of the death of his beloved brother, his only remaining relative. It discriminates between the predicament of the returning native and tourists to the locality who irritate the industrious local vicar as their 'arms' 'have a perpetual holiday'.[10] But Leonard still hankers for the local culture to which he half belongs, though it has become 'A place in which he could not bear to live',[11] reflecting what James Buzard has called 'the beginning of modernity ... a time when one stops belonging to a culture and can only *tour* it'.[12] He had become 'half a shepherd on the stormy seas', hearing

---

8 W. Wordsworth, *The Poetical Works of William Wordsworth*, ed. E. de Sélincourt, rev. H. Darbishire, 5 vols (Oxford: Clarendon, 1952–9), p. 262: 96.

9 K. Hanley, 'Wordsworth's "Region of the Peaceful Soul"', in K. Hanley and A. Milbank (eds), *From Lancaster to the Lakes, the Region in Literature* (Lancaster: Centre for North-West Regional Studies, 1992), pp. 12–14.

10 Wordsworth, *Poetical Works*, II 4: 7–9, 107.

11 Ibid., 13: 426

12 J. Buzard, *The Beaten Track: European Tourism, Literature and the Ways to Culture, 1899–1918* (Oxford: Oxford University Press, 1993), p. 26.

'The tones of waterfalls, and inland sounds/Of caves and trees', and waterfalls 'in the piping shrouds',[13] and experiencing a calenture which superimposed remembered scenes on a distant environment when he

> Below him, in the bosom of the deep,
> Saw mountains; saw the forms of sheep that grazed
> On verdant hills – with dwellings among trees,
> And shepherds clad in the same country grey
> Which he himself had worn.[14]

Wordsworth himself was a recurrent tourist. But wherever he wandered, he too took with him a similarly over-determining recollection, as in the sonnet, 'Composed Upon Westminster Bridge', where the inherent beauty of the London cityscape, with all its metropolitan and imperial prestige, is underwritten by bringing it into happy configuration with Lakes scenery: 'Never did sun more beautifully steep/In his first splendour, valley, rock, or hill'.[15] In his function as a guide to other tourists he attempted to bring them into closer relation to the world he had nearly lost, but had appreciatively regained, revealing to them a quality of existence which he could reveal to their outsiderness. His *A Guide Through the District of the Lakes in the North of England, with A Description of the Scenery, etc. For the Use of Tourists and Residents*, so entitled in its first separately published fifth edition of 1835, has been described as a major influence on holistic approaches to geography in the nineteenth century, 'covering not merely scenery but geology, landforms, climate, natural history, the society and economy of the inhabitants, their relationship with the landscape and their influence on the environment'. It was, in effect, 'a regional geography of the Lake District in which the interrelationships of a wide range of physical and human phenomena were examined'.[16] Beginning with helpful 'Directions and Information for the Tourist', by the third section he has drawn his incoming reader into sympathies with which to criticise other incomers who have become insensitive residents. Among the 'disfigurements' arising from bad taste, he describes how recent builders had adapted their architecture to the gaze of outsiders rather than integrating it with the customs of the indigenous community. They show a 'warping of the natural mind occasioned by a consciousness that, this country being an object of general admiration, every new house would be looked at and commented upon', so that 'Persons, who in Leicestershire or Northamptonshire would probably have built a modest dwelling like those of their sensible neighbours, have been turned out of their course', and their 'craving for prospect' causes them to build on 'the summits of naked hills in staring contrast

---

13   Wordsworth, *Poetical Works*, II: 46, 48–9, 47.

14   Ibid., II: 61–4.

15   Wordsworth, *Poetical Works*, III 38: 9–10

16   I. Whyte, 'Wordsworth's *Guide to the Lakes* and the geographical tradition', *Armitt Library Journal* 1 (1998), pp. 21–2.

to the snugness and privacy of the ancient houses'.[17] The notorious two letters and sonnets, *On the Projected Kendal and Windermere Railway*, which he published in *The Morning Post* in 1844 continued his anxious concern about the wrong kind of tourism – that the singular experience still available in the region should not be negated by an expanding tourist industry which engaged with it only by converting it into a consumer product.[18] The experience Wordsworth wished to preserve and extend was his urgent reconstruction of a pre-history of Lakes tourism. His *Guide* was conventionally addressed to 'the *Minds* of Persons of taste, and feeling for Landscape',[19] which the artists had first cultivated,[20] followed by Oxbridge dons' presentation of the Lakes in terms of the 'Italian' Picturesque. In 1786, William Gilpin, born at Scaleby near Carlisle, published his illustrated *Observations, relative chiefly to Picturesque Beauty, made in the year 1772, on Several Parts of England; especially the Mountains and Lakes of Cumberland, and Westmorland*, which became known as his "Northern Tour". He coined the phrase, 'Picturesque Beauty', defining rules to remodel landscapes for pictorial representation, but literal-minded conformity to these conventions laid the method open to satirical ridicule, as in *The Tour of Doctor Syntax, in Search of the Picturesque* (1809–1812), based on Thomas Rowlandson's illustrations (Figure 6.1). Nevertheless, in the opening decades of the nineteenth century, several collections of engraved picturesque landscape illustrations appeared in book form.

By the mid-eighteenth century, writers' interest in depicting their own local topography had developed markedly, and pre-Romantic writers deepened their responses in loco-descriptive poetry and the analysis of visual representation, scrutinising aesthetic terminology to extend its philosophical claims, as in the work of Burke, Uvedale Price and Payne Knight, which examined the principles of picturesque taste. But it was the guidebook writers who invited their readers to share the originating experience of the region itself. Thomas Gray was attracted to make a ten-day tour to this remote and unspoiled region which he recorded in his *Journal* of 1769. The year before its influential publication in an edition of his poems (1775), Thomas West produced *The Antiquities of Furness*, followed by the first guidebook to the Lakes, in 1778. The second and subsequent editions were considerably expanded by a local Kendal editor, William Cocklin, whose additions included an anthology of the accounts of previous literary visitors, including Gray. Spiritual resonances also began to emerge. John Wesley, who visited the Lakes

17 Wordsworth, *Guide to the Lakes*, pp. 73–4.

18 K. Hanley, 'In Wordsworth's Shadow: Ruskin and Nero-Romantic Ecologies', in G.K. Blank and M.K. Louis (eds), *Influence and Resistance in Nineteenth-Century English Poetry* (Basingstoke: Macmillan, 1993), pp. 228–31.

19 Wordsworth, *Guide to the Lakes*, p. 1.

20 K. Smith, *Early Prints of the Lake District* (Nelson: Hendon Publishing, 1973).

D⁰ SYNTAX TUMBLING INTO THE WATER.

Figure 6.1   'Dr Syntax Tumbling into the Water' by Thomas Rowlandson
Source: © The author's collection.

'every two or three years from 1748 until 1790, when he was 86', declared that he had found God 'nowhere more present than in the mountains of Cumberland'.[21]

Wordsworth's poetry similarly offered what in his *Guide*, though it was rooted in the picturesque tradition,[22] he described in more Romantic terms as a sense of the '*religio loci*'.[23] Besides the particularity of time and location in his works, there was a further, philosophical perspective in the representation of his experience which meant that 'To a greater degree than any other major English poet, Wordsworth was a poet of *place*'.[24] Yet more than for any other Romantic writer, Wordsworth's imagination depended on the actuality of real places. The philosophical world-view which informed his relation with place derived from a shift in Enlightenment thinking from a world beyond direct sensory perception to one *within* the operations of the human mind but itself based on sensory perception.[25] Enlightenment psychology fixed the entire field of knowledge within the scope of the imagination, as David Hume declared in 1739:

---

21   P. Hindle, *Roads and Tracks of the Lake District* (Milnthorpe: Cicerone Press, 1998), p. 76.

22   J.R. Nabholz, 'Wordsworth's *Guide to the Lakes* and the picturesque tradition', *Modern Philology* 61 (1964).

23   Wordsworth, *Guide to the Lakes*, p. 66.

24   D. Daiches and J. Flower, *Literary Landscapes of the British Isles: A Narrative Atlas* (London: Lund Humphries, 1979), p. 115.

25   N. Frye, '*The Drunken Boat*: the Revolutionary Element in Romanticism', in Frye (ed.), *Romanticism Reconsidered* (New York: Columbia University Press, 1963), p. 3.

Let us fix our attention out of ourselves as much as possible: Let us chance our imagination to the heavens, or to the utmost limits of the universe; we never really advance a step beyond ourselves, nor can conceive any kind of existence, but those perceptions, which have appeared in that narrow compass. This is the universe of the imagination, nor have we any idea but what is there produced.[26]

Wordsworth's imagination derived from that tradition of empirical realism, conflating the material and the fantastical, and discovering 'the region of its song' in the interactions of mind and nature:

Not in Utopia, subterraneous fields,
Or some secreted Island, Heaven knows where –
But in the very world which is the world
Of all of us.[27]

As Coleridge explained their revolutionary project, Wordsworth's early literary agenda was the awakening of his readers' minds to this extra potential within commonly available sensations: 'to give the charm of novelty to things of every day, and to excite a feeling analogous to the supernatural, by awakening the mind's attention from the lethargy of custom'.[28] Crucial for the link between material base and aesthetic superstructure was the realist psychology of Associationism, which informed Wordsworth's early poetry, providing the theoretical continuum of 'imaginative impressions',[29] from sensation to idea, so that literary associations carried the authenticity of actual experience.

'The Language of the Sense'

This imaginative insidedness is what Wordsworth offered to the traveller to the Lake District. Fiona Stafford has summarised the 'shifting balance between man and the environment, which accelerated rapidly in the eighteenth century', from inside to outside, and which Wordsworth's emphasis on imaginative experience confronts:

Instead of remaining active participants, people were gradually becoming observers, distanced from the land by changing social structures and economic, demographic, and employment patterns, together with advances in science,

---

26 D. Hume, 'A Treatise of Human Nature', in L.A. Selby-Bigge (ed.), *Hume's Treatise* (Oxford: Oxford University Press, 1965), I, Pt. 2, Sect. 6, pp. 67–8.

27 Wordsworth, *The Prelude* (1805), in J. Wordsworth, M.H. Abrams and S. Gill (eds), *The Prelude, 1799, 1805, 1850: The Norton Critical* Edition (New York and London, 979: Norton), 398: 10 723–26.

28 S.T. Coleridge, *Biographia Literaria* (London: Dent, 1977), p. 169.

29 Wordsworth, *Poetical Works*, I 330.

philosophy, technology, and agriculture. Modern man, increasingly cut off from his biological origins, was no longer an insider connected to the greater whole, while the external world became an object to view, chart, and quantify and an asset on which to capitalize.

With reference to Goldsmith's poem 'The Deserted Village', she writes that it 'was ... another name for the modern condition, a state of self-conscious division and inner conflict, which could strive to recover an ideal unity only through culture', and argues that 'Since the early nineteenth century, many readers who have never been to Cumbria, have still learned to love its imagined mountains and communities, sharing the sense of connectedness created by writing, because poetry invites readers to participate in its local truth'.[30]

Yet Wordsworth did not intend his poetry to be a substitute for a lost quality of experience, but rather to represent its continuing possibility in relation to an exemplary terrain. It is a crucial distinction between Wordsworth's sense of place and that more generally analysed by Nicola Watson, who argues that 'It is the internal workings of an author's works, buttressed by a particularized series of inter-texts, which produce place, not the other way around'.[31] Wordsworth certainly claimed that it was the other way around, and he repeatedly stated his recognition of the primacy of real experience:

> that such beauty, varying in the light
> Of living nature, cannot be portrayed
> By words, nor by the pencil's silent skill;
> But is the property of him alone
> Who hath beheld it, noted it with care,
> And in his mind recorded it with love.[32]

For him, 'Beauty, a living Presence of the earth' surpasses the 'most fair ideal Forms/Which craft of delicate spirits hath composed from earth's materials', [33] and that awareness led him to invite his readers to share his experience. Indeed, he accounted for a friend's 'impediments' in appreciating certain passages of his poetry as the result of lacking it:

> if a person has not been in the way of receiving these images, it is not likely that he can form such an adequate conception of them as will bring him into lively sympathy with the Poet. For instance one who has never heard the echoes of the

---

30   F. Stafford, *Local Attachments: The Province of Poetry* (Oxford: Oxford University Press, 2010), pp. 88, 90, 95.

31   N.J. Watson, *The Literary Tourist: Readers and Places in Romantic and Victorian Britain* (Houndmills: Palgrave Macmillan, 2006), p. 12.

32   Wordsworth, *Poetical Works*, V 303; IX, 512–17.

33   Ibid., 338: 754ff.

flying raven's voice in a mountainous Country, as described at the close of the
4th Book [of *The Excursion*] will not perhaps be able to relish that illustration.[34]

Those who had enjoyed the same sensations, he claimed, could also become poets:
'many are the Poets that are sown/By Nature ... Yet wanting the accomplishment
of verse'.[35] '*Silent*'[36] poets like his brother John participate in an unarticulated
potential which they may recognise in the writer's language, as Wordsworth
describes in an early draft for 'Michael', when the old man meets someone who
can put his own feelings into words:

> then, when he discoursed
> Of mountain sights, this untaught shepherd stood
> Before the man with whom he so convers'd
> And look'd at him as with a Poet's eye.[37]

Wordsworth's immersion in his milieu and the downplaying of his art has been
considered culturally limiting, for example by Ruskin, who in a late essay wrote
that 'Wordsworth is simply a Westmoreland peasant', whose 'studies of the
graceful and happy shepherd life of our lake country' are finally 'only ... the
mirror of an existent reality in many ways more beautiful than its picture'.[38] His
insistent materiality is what distinguishes him from the contemporary Scottish
poets of locality, Burns and Scott, who contributed to his own sense of localism
and whose relation to their home territory he tracked on his Scottish tour of 1803.
Wordsworth's attachments were different from theirs. While Burns had been the
first to become synonymous with his native district, Wordsworth observed that
'natural appearances rarely take a lead' [39] in his poetry, which is primarily based
on social feelings and tales. Scott's 'border vision', which Raymond Williams
describes as 'the special quality of writers who retain old ties while still achieving
a more detached view of their homes',[40] made him a more consciously literary
mediator of a separate and even exotic region, brokering a highly literary version of
traditional Scottish communities to a far-flung British and international readership.

The invitation to enter into the realist discipline of the Wordsworthian
imagination prompted Keats to follow in his footsteps when he visited the Lakes on

---

34   E. de Sélincourt (ed.), *Letters of William and Dorothy Wordsworth: The Middle Years*, rev. M. Moorman and A.G. Hill, 2 vols (Oxford: Clarendon, 1969–70), 'To Mrs Clarkson, Jan., 1815'; vol. 2, pp. 189–90.

35   Wordsworth, *Poetical Works*, V 10: I 77–8, 80.

36   Ibid., II 122: 80.

37   Ibid., II MS (b) 482.

38   E.T. Cook and A. Wedderburn (eds), *The Works of John Ruskin*, Library Edition, 39 vols (London: George Allen, 1903–12), vol. 34, pp. 318–19.

39   Quoted by Stafford, *Local Attachments*, p. 216.

40   Ibid., p. 149.

his northern tour of 1817, and, as he wrote to his brother, encountered the origin of what he had derived from reading Wordsworth: 'the tone, the coloring, the slate, the stone, the moss, the rock-weed; or if I may so say, the intellect, the countenance of such places'.[41] He discovered that indeed 'the ideas formed from his reading were inadequate to the reality of the scene before his eyes'.[42] But though Wordsworth's poetry depends on the originating experience it nevertheless has to attempt to convey it, and Keats recognised that creative representation could be closer to the originating experience than mere memory, as he wrote in a letter to a friend:

> Fancy indeed is less than a present palpable reality, but it is greater than remembrance – you would lift your eyes from Homer only to see close before you the real Isle of Tenedos – you would rather read Homer afterwards than remember yourself – one song of Burns is of more worth to you than all you could think for a whole year in his native country.[43]

Wordsworth's sequence of *Yarrow* poems, 'Yarrow Unvisited' (1807), 'Yarrow Visited' (1815) and 'Yarrow Revisited' (1835), closely teased out the interrelations between sensation and imagination in visiting places which have been highly pre-mediated, deferring finally to the permanence of place.[44] As the discursive superstructure of Wordsworth's poetry extended from the local attachment of the 1800 pastorals into an elevated discourse of moral nationalism in the Poems Dedicated to National Independence and Liberty, combining Burkean organicism with the high puritan idealism of Milton, and as his works subsequently became infused with Anglican quietism, he found the 'region of the peaceful self'[45] which he constructed as organically rooted in a specific place and community.

The translation from real experience into an emphasis on discursive representation was effected in his works of the early 1800s which are fascinated by indeterminacies between natural and human traces in the landscape. In verses he wrote when composing 'Michael', for example, he dwells on the simplest interventions – the cairn-piles by which travellers register their passing presence: 'some half o'ergrown/By the grey moss', left by travellers

> who in such places feeling there
> The grandeur of the earth have left inscrib'd
> Their epitaph which rain and snow
> And the strong wind have reverenced.[46]

---

41   Ibid., p. 249.
42   Ibid., p. 248.
43   Ibid., p. 251.
44   See Hanley, *Ruskin's Romantic Tours*, pp. 87–9.
45   *The Prelude* (1805) (see above, note 20), 396: 10 135.
46   Wordsworth, *Poetical Works* II 484: Fragment (d), 28–36.

Beyond individual correspondences the whole district became representational of regionalist ideology. Jonathan Bate has argued that the construction of what he calls 'critical regionalism' originated in Wordsworth's encounter with another English region on his return home to England after his visit to France during the lull in the Napoleonic Wars in 1802. Arriving at Dover he found himself in Kent, a county long associated with pre-Norman yoke protestant nationalism, and which offered images of locally based patriotic allegiances:

> The genealogy of Wordsworth's perception of Kent gives the 'Valley, near Dover' sonnet a different kind of patriotism from Burke's, a patriotism rooted not in evolving Westminster institutions ultimately of French origin, but in a tradition of local defence of liberty.

In that way, Bate sees Wordsworth's response as pre-formative of a federal view of Europe, which 'begins from the periphery, not the centre; it respects the distinctiveness of origins'.[47]

The crucial link between locality and national institutions was the Church of England, whose historical development is celebrated by what has been called Wordsworth's 'baptized imagination'[48] in the sequence of *Ecclesiastical Sketches* (1822) and *Ecclesiastical Sonnets* (1837), comprehending its spectrum of shifting positions, from Puritanism to Latitudinarianism to a High Church reverence for Catholic customs, and following its extended presence into North America. It was this tradition which underwrote his benevolent version of British imperialism, as he evokes it in his *Guide*:

> The chapel was the only edifice that presided over these dwellings, the supreme head of this pure Commonwealth; the members of which existed in the midst of a powerful empire like an ideal society ... and venerable was the transition, when a curious traveller, descending from the heart of the mountains, had come to some ancient manorial residence in the more open part of the Vales, which ... connected the almost visionary mountain republic he had been contemplating with the substantial frame of society as existing in the laws and constitution of a mighty empire.[49]

Wordsworth's religious interpretation of natural experience was immanent in previous constructions of the region's scenery. Thomas West was a Jesuit priest whose invention of the word 'station' for a picturesque vantage-point owed something to the contemplative reverence of the 'stations of the cross'. Though

---

47   J. Bate, *Critical Regionalism – Reflections on Wordsworth's Sonnets* (Ambleside: Armitt, 1997), pp. 13, 19.

48   J. Jones, *The Egotistical Sublime: a History of Wordsworth's Imagination* (London: Chatto and Windus, 1954).

49   Wordsworth, *Guide to the Lakes*, pp. 67–8.

Wordsworth's early poetry is that of 'at least a semi-atheist'[50] and pantheist, gradually and especially after the trauma of family deaths his responses became more creedal, and his poetry matured into an explicit Anglicanism. He came to see his poetry as expounding 'the Bible of the Universe',[51] and the different revisions of *The Prelude* became ever more spiritually explicit, influenced by the editing of his nephew, the Bishop of Lincoln.

In Wordsworth's Footsteps

In an 1802 review of Southey, the term 'a new sect' was used for what in an 1807 review of Wordsworth was described as 'a brotherhood of poets, who have haunted for some years about the Lakes in Cumberland', and whom the critic Francis Jeffrey and Byron later referred to as the 'lakers'. Subsequently they were sometimes referred to as the 'Lake Poets', a description which was eventually established in a collection of articles by Thomas De Quincey called *Recollections of the Lake Poets* when published as a volume in 1862.[52] The group included Coleridge, who went to live with his family at Greta Hall, Keswick, in 1800, and Robert Southey, whom Coleridge invited to join him there in 1803, promising Wordsworth as the chief among a 'society of men of intellect'.[53] Coleridge never returned to reside there after his therapeutic exile to Malta in 1804, while Southey, who became a prominent literary lion, preceding Wordsworth to the Poet Laureateship (1813–39), stayed until his death in 1843. Southey established a powerful literary presence in the district, but his writings seldom described the Lakes or related to the district materially, with the exception of his well-known literary *tour de force,* 'The Cataract of Lodore', which exuberantly mimics the movements of the falling water. The Lakes scene does figure in the background of Southey's political imagination – notably in the projection of the conservative utopianism which stems from his imagined dialogue with Sir Thomas More on the banks of Derwent Water in his *Colloquies on the Progress and Prospects of Society* (1822). But the great and good who visited Southey came to see the man, while participating in the customary diversions of Keswick, which was then the chief tourist centre of the Lakes: rural strolling and rambling, boating, theatricals and music.

Wordsworth offered different, and more rooted meanings. Other Romantic figures experienced their attraction, even entering into a kind of discipleship. Thomas De Quincey was the most significant; drawn by 'the deep magnet' of Wordsworth's poetry, he eventually settled in Grasmere in 1809, finally leaving Westmorland in 1828. In the revised edition of his *Confessions of an English*

---

50   S.T. Coleridge, *Collected Letters of Samuel Taylor Coleridge*, ed. E L. Griggs, 4 vols (Oxford: Clarendon, 1956–59), vol 1, p. 164.

51   E. de Sélincourt, *Letters*, 'To Mrs Clarkson', I, p. 78.

52   Hanley, 'Wordsworth's "Region of the Peaceful Soul"', pp. 24–5.

53   E. Dowden, *Southey* (London: Macmillan, 1879), p. 74.

Opium-Eater (1856), he looked back on his childhood imagination of the Lakes as he was growing up in Manchester:

> The southern region of that district, about eighteen or twenty miles long, which bears the name of Furness, figures in the eccentric geography of English law as a section of Lancashire... as Lancashire happened to be in my own county, I had from childhood, on the strength of this mere legal fiction, cherished as a mystic privilege, slender as a filament of air, some fraction of Denizenship in the fairy little domain of the English Lakes.[54]

When Charles Lamb brought his sister, Mary, to visit, they too thought they 'had got into Fairy Land'.[55]

The construction is that of an imaginatively privileged area, whose significance depended principally on Wordsworth's exemplary presence. The objective was literally to follow in his footsteps, and he eventually facilitated the practice for his readers by supplying information about actual locations in the notes for his poems which he dictated to Isabella Fenwick in 1843. Specific topographical guides to the poetry were later to emerge, particularly that by Wordsworth's Victorian editor, William Knight, *The English Lake District as Interpreted in the Poems of Wordsworth* (1878), describing itself as 'a guide to the Poems, rather than to the district; and to the district only in so far as it is reflected in, and interpreted by the Poems',[56] which were themselves considered the real guide. Knight subsequently provided the letter-press for 56 drawings by Harry Goodwin published as *Through the Wordsworth Country* (1887).[57] Though the illustrations served to fix the coincidence of poetry and place, they also widened the poetic allusions to encompass a more diffuse reference to typical landscape. Section XVI, 'Emma's Dell', for example, quotes the poem and identifies its setting as 'somewhere on the Easedale beck',[58] following the Fenwick note which places it loosely 'on the banks of the brook that runs through Easedale, which is in some parts of its course as wild and beautiful as brook can be'.[59] (Figure 6.2).

---

54   D. Masson (ed.), *The Collected Writings of Thomas De Quincey*, 16 vols (Edinburgh, 1862–71), vol. 1, pp. 73–5.

55   E.W. Marrs Jr (ed.), *The Letters of Charles and Mary Lamb* (Ithaca: Cornell University Press, 1976), Lamb to Thomas Manning; II, p. 68.

56   W. Knight, *The English Lake District as Interpreted in the Poems of Wordsworth* (Edinburgh: David Douglas, 1878), p. vii.

57   A modern version of this approach is D. McCracken, *Wordsworth and the Lake District: A Guide to Poems and Their Places* (Oxford: Oxford University Press, 1984). See also the section 'Topography and local associations to the poems', in K. Hanley, *An Annotated Bibliography of William Wordsworth* (Hemel Hempstead: Prentice Hall/ Harvester Wheatsheaf, 1995), pp. 43–5.

58   H. Goodwin, *Through the Wordsworth Country* (London: Swan Sonnenschein, 1877), p. 77.

59   Wordsworth, *Poetical Works*, II 486.

Figure 6.2   'Emma's Dell', in Harry Goodwin's *Through the Wordsworth Country*, 1887
Source: © The author's collection.

As Knight acknowledges, Wordsworth was unconcerned with pinning down the exact places – 'Oh yes; that, or any other that will suit' – and sometimes persisted in employing the picturesque license of assembling details from various places: Wordsworth had told a friend that copying details missed the point of good nature poetry which should be based on the mind's interactivity: 'That which remained, the picture surviving in his mind, *would have presented the ideal and essential truth of the scene, and done so in a large part, by discarding much which, though in itself striking was not characteristic*'. His words conveyed his generic experience of real places, as Knight claimed: 'Homer can be understood without a visit to the Troad, or the Aegean; but the power of Wordsworth cannot be fully known by one who is a stranger to Westmoreland', so that the representational power of those places could be fully appreciated peculiarly through Wordsworth's mediation: 'It is ground which [his poems] have *made* classic … interpreting the scenery as it now is, by laying hold of its inner meaning, its perennial spirit, and embodying its deep underlying significance, in imaginative forms of unparalleled grace and power'.[60] A number of subsequent studies followed Knight's lead, and the turn to complementary visual representation became established. Knight welcomed the additional medium of Goodwin's illustrations, which had 'reproduced ... features [of the Wordsworth Country] and made them permanent', as a way of retaining the vestige of materiality, of the historical base of a realist encounter which

> will cease to be realizable even there, unless the vale of Grasmere remains a place of seclusion, unviolated by the roar of machinery and the din of traffic. The blasting of quarries, and the works of the water-engineers, have already done it grievous wrong ... if ever the steam-car enters the vale of Grasmere, the lover of repose will have to seek it elsewhere, and the charm of that indescribable region will be only a memory of the past.[61]

An issue behind Knight's promotion of visual mediation was his recognition that as the poet's texts became ever more widely disseminated, the topographical references had to be kept up for a readership which mostly could only connect with the real places in that way:

> All experience shows, moreover, that posterity takes a great and growing interest in exact topographical illustrations of the works of great authors. I need only refer to the labour recently bestowed upon localities associated with Shakespeare, and with Burns, and to the success which has attended it.[62]

At the same time, however, an appreciation of Wordsworth's representation of place was becoming more generally available for actual experience through the popular access enabled by increased mobility in the railway age. The development was ironical in view of the Romantic misgivings about intrusive tourism, but not all Wordsworth's friends agreed with him in resisting the incursions of modernism. Harriet Martineau advocated the benefits of the anti-Romantic tendency, offering an unvarnished description of her local community:

> Her *Guide to the English Lakes*, first published in 1855, showed more interest in the availability of baths in hotels and in the organization of horse-drawn transport than in romantic evocations of sublime landscape, and her matter-of-fact tone extended to the peasantry, who were presented as primitive, ignorant, tongue-tied, credulous and drink-sodden.[63]

---

60 Knight, *English Lake District*, pp. viii–xii.
61 Goodwin, *Through the Wordsworth Country*, p. xvii.
62 Ibid., p. xv.
63 J.K. Walton, 'Canon Rawnsley and the English Lake District', *Armitt Library Journal* 1 (1998), p. 3.

She advocated the spreading of the commercial spirit to remote hamlets. Middle-class tourism to the Lakes, concentrated around the two principal centres of Keswick and the east bank of Windermere, had consumed the picturesque aesthetics as part of a holiday agenda of recreational diversions and therapeutic motives from the beginning of the century.[64] For many visitors, the Lake Poets themselves attracted more attention than the sites of their poems, as the eleven-year old Ruskin, in his verse journal of their visit in 1830, *Iteriad; or Three Weeks Among the Lakes*, with topographical notes, describes his family attending Rydal Chapel and Crosthwaite Church to catch a glimpse of 'old Mr Wordsworth at chapel of Rydal' and Southey's 'dark lightning eye [which] made him seem half-inspired'[65] in Crosthwaite Church. During the 1830s Wordsworth's later home, Rydal Mount, increasingly became the object of sight-seers' curiosity, of the sort that had previously formed around Scott and Abbotsford. William Howitt's popular *Homes and Haunts of the Most Eminent British Poets* (1847) confirmed the market for those who wished to imagine and reconstruct the habitats of favourite writers. All Wordsworth's homes are mentioned, but Dove Cottage hardly at all, being referred to simply as one of the 'two or three different homes' in which he resided at Grasmere. Instead, Wordsworth is primarily the poet of Rydal Mount, 'where he has now lived for more than thirty years' in a home which is 'as perfectly poetical in its location and environs as any poet could possibly conceive in his brightest moments of inspiration'.[66]

After Wordsworth

Over time, the Lakes became the centre of an interconnected literary culture with a core of resident writers, all associated with the district and with Wordsworth personally. The retinue of Romantic and Victorian writers included De Quincey, who rented Fox Ghyll from 1821–25, John Wilson (alias 'Christopher North'), who built Elleray on his estate on Windermere, Harriet Martineau who built The Knoll in Ambleside, Felicia Hemans who rented part of Dove Nest on Windermere, and Thomas Arnold and his family, including Matthew, who holidayed at Fox How on Loughrigg from 1836–42.[67] Ruskin came later, in 1871, and took root at Brantwood, at a point on Coniston Water to which Wordsworth's *Guide* had drawn attention. The process by which this regional network became nationally institutionalised was recorded and significantly facilitated by Canon Rawnsley, who enthusiastically assimilated the Anglican Wordsworth as the

---

64   Hanley, *Ruskin's Romantic Tours*, pp. 37–68.

65   Cook and Wedderburn, *John Ruskin*, 4 113: 613; 2 60:299.

66   W. Howitt, *Homes and Haunts of the Most Eminent British Poets*, 2 vols (London: R. Bentley, 1847), vol 2, pp. 268, 286.

67   P. Whitehead, *They Came to the Lakes: A Guide to the Homes of Famous Poets & Writers* (Clapham: Dalesman, 1966).

presiding spirit of the Victorian Lakes, notably in his two volumes on *Literary Associations of the English Lakes* (1893). His own *Sonnets at the English Lakes* (1886) retrace a familiar poetic map with the wise moral sentiment of 'Nature's Gospel': 'Here find the still communion that avails/To fire imagination almost dumb'.[68] Rawnsley's *Ruskin and the English Lakes* (1901) contains two chapters on 'Ruskin and Wordsworth', highlighting the continuities between both writers: 'Wordsworth's experience was Ruskin's experience also', and observing that they shared a reverence 'for Westmoreland peasantry'.[69] In their similar letters against the railways both refuted the value system of industrial capitalism,[70] advocating unalienated work practices which Wordsworth had described in such sonnets of 1803 as 'These times touch monied worldlings with dismay',[71] and Ruskin in *Unto This Last* (1862). Rawnsley dwelt on their common 'belief in the educative power of scenery',[72] as Ruskin urged it recurrently, for example in 'Of Queens' Gardens' in *Sesame and Lilies* (1865) recommending the upbringing of young girls according to the natural joy described in Wordsworth's 'Three Years She Grew in Sun and Shower': 'And vital feelings of delight/Shall rear her form to stately height'.[73]

Rawnsley concluded *Ruskin and the English Lakes* with the image of two graves, Wordsworth's at Grasmere since 1850, and Ruskin's at Coniston in 1900, which he described as 'consecrated' shrines: 'and pilgrims ... will think of the double debt we owe'.[74] They represented the absorption of presence into place, to which M.J.B. Baddeley's popular late Victorian *Thorough Guide to the Lake District* (1880), adjured the visitor to pay his dues: 'the first thing one does in Grasmere is to visit Wordsworth's grave'. For Rawnsley these sites had become a 'National Resting-ground', where the writers' 'spirit words'[75] had become re-embodied in the landscape for future generations. He was unabashed in promoting the national ethos of Romantic regionalism as more positive than simply an 'elegiac record of humanity's sense of alienation from its original habitat in an irrecoverable, pre-capitalist world'.[76] His account of 'Daffodil Day at Cockermouth' in *Life and Nature at the English Lakes* (1902) is the culmination of his description of local customs and festivities. He describes a new commemoration in the public park, two years after a fountain had been constructed there in memory of Dorothy and

---

68   H.D. Rawnsley, *Sonnets at the English Lakes*, 2nd edn (London: Longmans, Green and Co., 1892), p. 3.

69   H.D. Rawnsley, *Ruskin and the English Lakes* (Glasgow: J. MacLehose, 1901), pp. 168–9.

70   See Hanley, 'In Wordsworth's Shadow', pp. 228–31.

71   Wordsworth, *Poetical Works*, III 119: 1.

72   Rawnsley, *Ruskin and the English Lakes*, p. 172.

73   Wordsworth, *Poetical Works*, II 214: 31–2

74   Rawnsley, *Ruskin and the English Lakes,*. p. 187.

75   Ibid., pp. 187–8.

76   Ibid., pp. 21–2.

William's birth and childhood, at which a band of boys from the Industrial Home led the march of the assembled local schoolchildren to place daffodils in the basin. Proceedings included an address by the Provost of Eton and the volume ends: 'my mind wandered away to the Klondike gold fields, and back to the great heap of blossoming gold there in the fountain basin. I could not but hope that the spirit which Wordsworth taught us, of love for a simpler life, should be reborn into the heart of this century'.[77]

Howitt's account of Rydal Mount was of the habitat of a still living poet. When Stopford Brooke visited Wordsworth's other home, Dove Cottage, in 1912, the building which was to become the iconic home not just of Wordsworth, his family and literary associates, but also of English poetry, his imaginative construction of its potential cultural content depended on the retention of literary links which were still palpable: 'There is no place ... which has so many thoughts and memories belonging to our poetry ... And it is almost untouched'.[78] The haunting associations had become more substantial after Knight's publication of Dorothy Wordsworth's complete journals, including the Grasmere Journal, in 1897, with its minute and sensitive natural descriptions of the area, and subsequently the three-volume edition of *The Letters of the Wordsworth Family* (1907). Stopford Brooke produced a shilling booklet in which the visit was described, as an appeal for the place to be 'secured for posterity' for 'reasons drawn from the associations, personal, literary, and moral, which gather round this cottage, from the poetic sentiment which abides in it, and from its preciousness to English-speaking men and women over the whole world'.[79] Helped by Knight, he was soliciting a 'national subscription' to preserve 'a memorial of Wordsworth ... for the pleasure and good of the English race' following the model of Shakespeare's birthplace at Stratford: 'Why should we not try and secure it, as Shakspeare's [sic] birthplace is secured, for the eternal possession of those who love English poetry all over the world? ... both in these islands, in our colonies, and among our brothers and sisters in America'.[80]

As Julia Thomas has argued, the auction of Shakespeare's birthplace in Stratford in 1847, by conflating bard and birthplace, had given rise to the 'Shakespeare Industry' or 'Shakespearean exchange economy' of Stratford as a tourist mecca. On the one hand, the birthplace was felt to retain an irreplaceable connection with real experience, but on the other the cultural meaning was considered detachable and available in multiple forms of representation: indeed, an exact replica was set up in the Surrey Zoological Gardens. The rhetoric of the Stratford appeal – a 'hallowed shrine' and a 'place of pilgrimage' – claimed a sacred cultural

---

77   H.D. Rawnsley, *Life and Nature at the English Lakes* (Glasgow: J. MacLehose, 1902), p. 271.

78   Howitt, *Homes and Haunts*, p. 14.

79   S. Brooke, *Dove Cottage: Wordsworth's Home from 1800–1808* (London: Macmillan, 1913), p. 21.

80   Ibid., p. 15.

significance which attempted to establish its 'otherness from the market'[81] even while it was effectively constructing a localised market. The Wordsworth industry was to follow the same route, as the real places and their associations became assimilated into a tourist package of heritage attractions, while reconnecting the region with the sources of its literary representation has become an increasingly specialised cultural and educational practice.

81   J. Thomas, 'Bidding for the Bard: Shakespeare, the Victorians, and the auction of the birthplace', *Nineteenth-Century Contexts* 30, 3 (September 2008), pp. 216, 219.

Chapter 7

'Inhabited by Strangers': Tourism and the Lake District Villa

Adam Menuge

Rash Assault

In 1769, while gathering material for his *Six Months Tour through the North of England*, Arthur Young approached Windermere from Kendal, and hiring a boat from the innkeeper at Bowness was rowed out to sample 'the extreme beauty of the lake' – particularly the situation of the largest island on it. '*The Island*', Young observes,

> contains between thirty and forty acres of land, and I cannot but think it the sweetest spot, and full of the greatest capabilities, of any forty acres in the king's dominions. The view from the south end is very fine; the lake presents a most noble sheet of water stretching away for several miles, and bounded in front by distant mountains; the shoars beautifully indented by promontories covered with wood, and jetting into the water in the most picturesque stile imaginable.[1]

And he goes on to describe, in minute detail, the views to be enjoyed from the island in every direction.

Young (1741–1820) was by no means a typical tourist. He was motivated primarily by the desire to examine the progress of his countrymen in the great work of agricultural improvement. But he was a man of sufficient culture to relish the incidental opportunities afforded by his tour to view the increasingly celebrated scenery of the Lake District, recording his impressions, not in the main body of his text, but in two immense footnotes. His use of the word 'capabilities' reminds us, significantly, that in the 1760s it was still unusual for a conception of landscape to be divorced from considerations of its actual or potential benefits, either as a basis for landowning wealth and the enjoyment of leisure, or as the sustenance of the contented tenantry and peasantry on which depended the safety of the political establishment. More particularly, it calls to mind the greatest landscape designer of the generation before Young's, Lancelot 'Capability' Brown (1716–1783), and his transformations of numerous country house parks.

---

1   A. Young, *A Six Months Tour through the North of England*, 2nd edn, 4 vols (London: W. Strahan & W. Nicholl, 1771), vol. 3, pp. 138n–9n.

Young, though he is now accounted one of the pioneers of Lakes tourism, was not the first to savour the sights of Windermere. The fact that the innkeeper 'keeps a boat, and can always provide rowers for any company that comes' is a measure of how well established tourism had already become.[2] Nor does Young ever seem to have lacked someone locally knowledgeable to show off the best viewpoints. Indeed the island (then known as Longholm), on which a house (Holm House) had stood since at least the mid-seventeenth century,[3] must for long have been enjoyed by its minor gentry owners, the Philipsons, as some kind of retreat, and had already attracted the attentions of 'off-comers' before Young's visit. In 1749 it was acquired by John Floyer, described in title deeds[4] as of Longdon, Staffordshire, and the following year Dr Pococke noted, in the course of his northern tour, that 'Mr Floyer, a Roman Catholick, is lately come from Coventry to live in a retired way'.[5] In 1752, for reasons unknown, Floyer sold the island to Thomas Barlow Esq. of Salford, Lancashire. Nicolson and Burn noted that the island was famous for 'a remarkable echo; and for hearing the same in perfection Mr Barlow provided two small cannon'.[6] The discharging of cannon and other firearms in this manner was a diversion popular among early tourists,[7] and it is clear that Barlow's enjoyment of his island was essentially recreational. The property subsequently descended through the Barlow family and on 25 March 1772 it was conveyed by Robert and Thomas Barlow to Thomas English.[8] By 1774 English, of St Pancras, London, was building an ambitious classical villa to designs by the metropolitan architect John Plaw (*c.*1745–1820), but his enjoyment of the house was brief: by 1781 it had been bought on behalf of Isabella Curwen of Workington Hall, in whose family it remained until 1991.[9]

The story of Belle Isle, as the villa built by English came to be known, and of its reception by contemporaries has been told at length elsewhere.[10] A modern

---

2   Ibid., pp. 137n–8n.

3   J.M. Ewbank (ed.), *The Collections of the Rev. Thos. Machell, Chaplain to King Charles II, towards a History of the Barony of Kendal*, Cumberland & Westmorland Archaeol. & Antiq. Soc. Extra Series 19 (Kendal, 1963), pp. 87–88.

4   Cumbria Record Office (Whitehaven) (henceforth CRO(W)), D/Cu, compartment 15.

5   J.J. Cartwright (ed.), *The Travels through England of Dr Richard Pococke*, 2 vols (Camden Society, new ser. 42 (1888) & 43 (1889)), vol. 1, p. 42.

6   J. Nicolson and R. Burn, *The History and Antiquities of the Counties of Westmorland and Cumberland*, 2 vols (London: W. Strahan & T. Cadell, 1777), vol. 1, p. 186.

7   See, for example, J. Clarke, *A Survey of the Lakes of Cumberland, Westmorland, and Lancashire* (London: J. Clarke, 1787), p. 23.

8   This conveyance does not appear to survive, but is mentioned in a recitation of title contained in a draft indenture of 1781 (CRO(W), D/Cu, compartment 15).

9   CRO(W), D/Cu, compartment 15; Curwen monument in St Martin's, Bowness.

10 See E.W. Hodge, 'Belle Isle', *Country Life* (3 and 10 August 1940); J.M. Robinson, 'Belle Isle', *The Connoisseur* (March 1981), pp. 234–7; A. Menuge, 'Belle Isle, Windermere, Cumbria', RCHME Historic Building Report, 1997; A. Menuge, 'Belle

environmental sensibility will perhaps understand the hostility it attracted from contemporaries. Already, by the early 1770s, the views from the island, and the view across Windermere from the slopes above the Ferry Inn, in which the island featured prominently, were widely admired. They were only a few years away from being formalised as the first, second and third viewing 'stations' of Fr Thomas West (c.1720–1779), whose influential *Guide to the Lakes* was first published in 1778,[11] and around 1799 the elevated viewpoint received a further imprimatur in the form of the belvedere (discussed in Sarah Rutherford's chapter) built by Revd William Braithwaite of Satterhow. To build a large villa in the midst of such a public arena was simultaneously to acknowledge the esteem in which it was held and abruptly to change its signification. Plaw's depiction of the house in the frontispiece to his collection of designs, *Rural Architecture* (1785), makes clear that it was conceived as an ornament to an Arcadian landscape in the manner of an antique temple in a painting by Claude.[12] It revealed, however, the contradiction at the heart of the villa-building enterprise: how to enjoy Nature in her 'beautiful unaffected state' (as West put it)[13] without eroding precisely that same precious quality by importing the intrusive trappings of everyday culture. William Hutchinson, who saw the works in progress, put it thus:

> The few natural beauties of this island are wounded and distorted by some ugly rows of firs set in right lines, and by the works now carrying on by Mr English, the proprietor, who is laying out gardens on a square plan, building fruit walls, and preparing to erect a mansion-house. – The want of taste is a misfortune too often attending the architect; – the romantic site of this place, on so noble a lake, and surrounded with such scenes, requires the finest imagination and most finish'd judgment to design the plan of an edifice and pleasure grounds; – but instead of that, to see a Dutch Burgomaster's palace arise, and a cabbage garth extend its bosom to the east, squared and cut out at right angles, is so offensive to the traveller's eye, that he runs away in disgust.[14]

Isle and the Invention of the Lake District', in M.F. Hopkinson (ed.), *Town and Country: Contemporary Issues at the Rural/Urban Interface* (York: PLACE Research Centre, 2000), pp. 71–82; and P. Leach, 'The house with a view in late eighteenth-century England: a preliminary inquiry', *Georgian Group Journal* 16 (2008), pp. 117–31.

11   T. West, *A Guide to the Lakes: dedicated to the lovers of landscape studies, and to all who have visited, or intend to visit the Lakes in Cumberland, Westmorland and Lancashire. By the author of the Antiquities of Furness* (London: Richardson & Urquhart and Kendal: W. Pennington, 1778).

12   J. Plaw, *Rural Architecture: or, designs from the simple cottage to the decorated villa* (London: J. Taylor, 1785), frontispiece.

13   West, *A Guide to the Lakes* (1778 edn), p. 63.

14   W. Hutchinson, *An Excursion to the Lakes in Westmoreland and Cumberland; with a Tour through part of the Northern Counties in the years 1773 and 1774* (London: J. Wilkie, 1776), pp. 187–8.

English's architect, John Plaw, worked principally in London and the south-
east, but it is clear that his building was a response to very particular local
circumstances. Belle Isle (Figure 7.1) is striking in having a circular plan, rising
from a concealed square basement area and topped by a huge dome. The rooms
are arranged around the central stair hall, which is top-lit by a circular lantern
crowning the dome. All the fireplaces are located on the walls enclosing the stair
compartment, enabling the flues to be carried up invisibly through the masonry
piers of the lantern, and keeping the circumferential wall clear of obstructions.
Here a portico beckoned eastwards to Bowness – for many tourists their first
calling point in the Lakes. Large Venetian and tripartite windows lighting the
principal rooms faced north and south, up and down the lake, and west to the
steeply rising Claife Heights, while smaller windows coincided with the second
rank of compass points. Clearly the intention was to have as full a command of
the available prospects as possible – a panopticon, no less.

Figure 7.1   Belle Isle, Windermere, photographed in 1940
Source: © *Country Life*.

The example of Thomas English was quickly followed by Joseph Pocklington (1736–1817), the wealthy son of a Newark banker. Pocklington bought the largest island on Derwent Water, near Keswick, where in 1778 he built a small villa to his own designs, doubling its size very shortly afterwards. He also set about ornamenting the island with a range of gimcrack curiosities, including a mock-church and fort, and a 'Druids' Temple'. Pocklington, who was not a prey to false modesty, renamed his island Pocklington's Island and with Peter Crosthwaite (1735–1808), the Keswick hydrographer and museum keeper, he promoted a series of regattas on the lake, in which his island 'defences' were attacked and defended by scores of men hired purposely for the entertainment.[15] As an architect he had no special talent, and as a showman he divided opinion, but as an enthusiast for lake scenery, who translated impulse into action, he had no equal. Within a very few years he had also built Finkle Street House (1784) at Portinscale, overlooking the lake from north-west, and Barrow House (1787) on the eastern side of the lake, with a noted cascade just to its rear. By dint of multiplying his residences – and flitting between them – he achieved something like the diversity of views assembled in a single house at Belle Isle.

In architectural terms there were no direct heirs to Belle Isle in the Lake District, and the only two eligible islands having been appropriated, no more island houses were built. In the 1770s the overwhelming response to private enjoyment had been public displeasure, but during the 1780s and 1790s villas nevertheless began to proliferate, hastened at least in part by the fact that from 1792 the Continent was effectively closed to British tourists.

Leisure and the Rise of the Villa

At this point it will be helpful to define more precisely what is meant by the term 'villa'. As so often, historical usage is deplorably inconsistent. The new houses which sprang up for those bent upon enjoying the Lake District at leisure between 1780 and 1914 were variously called villas, seats, residences, mansions and cottages, labels in which issues of status weigh more heavily than formal or functional criteria. Many house names incorporated similarly value-laden terms ('cottage', 'hall', 'lodge'); a few affected a specious antiquity ('tower', 'castle'); and a handful used the word 'villa'. They varied widely in size, form and style, and

---

15   For Pocklington and Crosthwaite, see A. Taylor, 'Compulsive Lakeland builder: Joseph Pocklington', *Country Life* (5 Sept 1985), and ibid., '"More Vile Taste ...": the Pocklington Brothers and their buildings', *Country Life* (1 May 1986); P. Crosthwaite, *Seven Maps of the Lakes* (Keswick: P. Crosthwaite, 1783); A. Hankinson, *The Regatta Men* (Milnthorpe: Cicerone Press, 1988); P. Brears, 'Commercial museums of eighteenth-century Cumbria: the Crosthwaite, Hutton and Todhunter collections', *Journal of the History of Collections* 4, 1 (1992), pp. 107–26; and M.E. Brown, *A Man of No Taste Whatsoever* (Milton Keynes: AuthorHouse, 2010).

there is only the loosest correlation between these variables and the labels applied by contemporaries to individual houses. The term 'villa' emerged in the early eighteenth century to describe a series of elite – often aristocratic – residences close to London, functioning as a retreat from the fatigue and ennui of the London social season, with its attendant business, jobbery and litigation. It quickly became a convenient label for suburban, peri-urban and rural houses effecting a separation of business and leisure among the rising, and essentially urban, commercial, professional and manufacturing middle class. In the nineteenth century the social signification of the term slipped further, embracing larger semi-detached houses and even (though such usage may be ridiculed) some terraces. But if the term was open to abuse, the central core of meaning survives intact. At its heart the villa represents an escape – real or imagined – from the busy world, and a vision, however dim, of the earthly paradise, expressed through relative seclusion, a striving after architectural refinement and a calculated response to landscape.

Tourism and the villa have this in common, then, that they are both key manifestations of the increase in leisure – more precisely, the consumption of leisure – which has helped to define the modern world. It is no accident that the area of rural England most inextricably bound up with the growth of domestic tourism in the second half of the eighteenth century and throughout the nineteenth century is also the area with the greatest concentration of rural villas, nor that they are sited in conformity with the same principles as informed the choice of the various viewing 'stations'. Yet the phenomenon of villa-building has hitherto been little studied, either in its own right or in the context of developing tourism. My interest in these themes was first stimulated by the prolonged interdisciplinary collaboration in the 1980s between the literary scholar Robert Woof of the Wordsworth Trust, the art historian John Murdoch of the Victoria and Albert Museum, and the bibliographer Peter Bicknell, and the resulting series of exhibitions and papers. The principal subject of these was the artistic and literary response to the Lake District landscape and the growth of tourism which partly fuelled and partly fed upon it; within this wider context the villa assumed, from time to time, a modest degree of prominence.[16] Mainstream architectural history scholarship has favoured the elucidation of design provenance, often focusing

---

16   P. Bicknell and R. Woof (eds), *The Discovery of the Lake District 1750–1810: A context for Wordsworth* (Grasmere: Trustees of Dove Cottage, 1982); P. Bicknell and R. Woof (eds), *The Lake District Discovered 1810–1850: The Artists, the Tourists, and Wordsworth* (Grasmere: Trustees of Dove Cottage, 1983); J. Murdoch (ed.), *The Discovery of the Lake District: A Northern Arcadia and its Uses* (London: Victoria & Albert Museum, 1984); J. Murdoch (ed.), *The Lake District: a sort of national property. Papers presented to a symposium held at the Victoria & Albert Museum, 20–22 October 1984* (Cheltenham: Countryside Commission and London: V&A, 1986); D. Thomason and R. Woof (eds), *Derwentwater: The Vale of Elysium* (Grasmere: Trustees of Dove Cottage, 1986); and J. Murdoch, 'A Villa in Arcadia', in S. Pugh (ed.), *Reading Landscape: country – city – capital* (Manchester and New York: Manchester University Press, 1990), pp. 121–44.

on single buildings or small samples selected on the basis of their architectural pedigree rather than on their relationship to broader social trends or landscapes.[17] This essay – part of a larger work in progress – attempts to draw out more fully the complex and still developing relationship between tourism and the villa, with a particular emphasis on the period 1770–1850.

Distribution and Evolutionary Patterns

The distribution of villas across the Lake District was, and is, far from even. At a very basic level, a high premium was placed upon a lake view – and preferably a lake frontage – and consequently few villas were located in valleys (such as Eskdale and St John's Vale) lacking a sizeable body of water. It is nevertheless striking that as many as 90 per cent of the villas built between 1770 and 1914 stand within sight of just seven of the region's fifteen major lakes.[18] Aesthetic judgements favouring lakes with particular characteristics account for some of this bias, but the geography of the Lake District also had a pronounced effect. Most tourists approached from the south or the east, by road via Lancashire or one of several transpennine routes, or by sea from Liverpool and (more briefly) Fleetwood. Thereafter their course was constrained by severe deficiencies in the road network: the only satisfactory east-west route through the Lake District was that between Penrith and Cockermouth, well to the north of the main western valleys, and the more southerly coastal route, with its repeated tidal crossings, was both circuitous and slow (though the first crossing, from Hest Bank near Lancaster to a point near Grange-over-Sands, was a popular approach recommended by Wordsworth among others). For a variety of reasons, then, the western valleys were unattractive to most potential villa-builders, and remained so even after the railway arrived. The proximity of markets was an equally important consideration, since most villa builders expected their creature comforts to be satisfied. A number of long-established markets were scattered around the fringes of the Lake District,

---

17   Rare exceptions are J.M. Crook, 'Privilege and the Picturesque: New Money in the Lake District, 1774–1914', in *The Rise of the* Nouveaux Riches: *Style and Status in Victorian and Edwardian Architecture* (London: J. Murray, 1999), pp. 79–100, and G. Sheeran, *Brass Castles: West Yorkshire New Rich and their Houses 1800-1914* (Stroud: Tempus, 2006). Useful general studies include J.A. Ackerman, *The Villa: Form and Ideology of Country Houses* (London: Thames & Hudson, 1990) and D. Arnold (ed.), *The Georgian Villa* (Stroud: Alan Sutton, 1996). M. Girouard, 'What is a Villa?', in *Town and Country* (New Haven and London: Yale University Press, 1992), pp. 235–8, is a succinct introduction.

18   To arrive at this estimate I have excluded the numerically significant, but somewhat separate, group of villas bordering Morecambe Bay. The seven lakes are Bassenthwaite, Coniston Water, Derwent Water, Esthwaite Water, Grasmere, Ullswater and Windermere. The figure of fifteen is arrived at by adding Buttermere, Crummock Water, Ennerdale Water, Hawes Water, Lowes Water, Rydal Water, Thirlmere and Wastwater.

notably at Penrith, Cockermouth, Broughton and Ulverston, but it was the small markets of Hawkshead, Bowness, Ambleside and Keswick that exerted the most powerful influence on the location of villas.

The villas built before about 1800 congregated, with very few exceptions, along the eastern shore and northern tip of Windermere, the northern end of Derwent Water and the lower reaches of Ullswater. The conspicuous cluster of villas around Windermere attracted particular attention. The sixth edition of Thomas West's *Guide to the Lakes*, published in 1796, included the following footnote:

> On the banks of Windermere-water, have been lately built, or are now building, a number of elegant villas; by Mr. Law, at Braithay; Miss Pritchard, Croft-Lodge, Clappersgate; Mr. Harrison, above Ambleside; Mrs. Taylor, Cottage, Ambleside, the Bishop of Llandaff, at Calgarth; Mrs Taylor, Bells-Field, near Bowness; Sir John Legard Bart. near Storrs; Mr. Dixon, Fell-Foot; and others. These works of art, most of which are done in stiles suitable to their situations, give an air of great consequence to the country, and, with the surrounding natural beauties, have lately made this neighbourhood, and particularly about Ambleside, a place of the greatest resort.[19]

West had inaugurated the guidebook writer's habit of drawing attention to notable villas with his brief critique of Belle Isle in 1778. But guidebook writers (who could not afford to offend a substantial section of the book-buying public) had reason to be circumspect, whereas journal and letter writers were often outspoken. It is for this reason, no doubt, that West's posthumous editor, William Cockin (1736–1801), in the third edition of the *Guide*, remonstrated with his author for his mild criticism of Belle Isle.[20] Since the work had first appeared Belle Isle had changed hands; it was no longer the frivolous toy of a geographical and social outsider who could be goaded with impunity, but the holiday villa of the Curwens, local magnates whose large estates and industrial wealth demanded respect.[21]

---

19   T. West, *A Guide to the Lakes in Cumberland, Westmorland, and Lancashire*, 6th edn (London: W. Richardson, J. Robson and W. Clarke and Kendal: W. Pennington, 1796), pp. 70n–71n. The identity of some of the houses is not immediately apparent. Mr Harrison's is Scale How, Ambleside, latterly part of Charlotte Mason College and its successors; Mrs Taylor's 'Cottage' is Iveing Cottage on the Old Lake Road, Ambleside, while 'Bells-field' should not be confused with the villa (and now hotel) of the same name – it is in fact Belfield, Bowness.

20   West, *A Guide to the Lakes* (1778 edn), pp. 63–64; ibid., 3rd edn (London: B. Law, Richardson & Urquhart and J. Robson and Kendal: W. Pennington, 1784), pp. 60n–61n.

21   In 1781 Belle Isle was purchased for Isabella Curwen of Workington Hall, then a minor. The following year she married her cousin and erstwhile trustee John Christian (1756–1828) of Unerigg. He assumed the Curwen name in 1790, became a noted exponent of agricultural improvement and commercial afforestation, and was elected M.P. successively for Carlisle and Cumberland.

The new villas of the late eighteenth and early nineteenth century varied considerably in form. At one extreme there were opulent villas in the manner of small Palladian country houses at Calgarth Park (1789) and Brathay Hall (built between 1788 and 1796), both on Windermere. More common were the modest three-bay houses, usually given extra gravitas by small flanking wings, such as Dove Nest (before 1788) near Ambleside, Lake Bank (1812) overlooking Esthwaite Water and Highfield Cottage, Blawith (Figure 7.2). Smaller still were the single-storey 'cottages' of Lord Gordon at Derwent Bay (*c.*1784), near Portinscale, and John Wilson at Elleray (1808), Windermere. The builders of the new villas were equally diverse. A few, such as Lord Gordon and Sir John Legard, the first builder of Storrs Hall (*c.*1795), Windermere, belonged to the traditional landed and titled elite. A few more were wealthy clerics, such as Richard Watson, absentee Bishop of Llandaff, who built Calgarth Park. Inevitably London threw up its share of villa builders, and other distant parts of Britain were not unrepresented, but many more villas were the product of the new commercial wealth generated in Midland and Northern England. In the late eighteenth and early nineteenth centuries they came especially from Liverpool and other north-western sea-ports such as Ulverston and Whitehaven, but over time this commercial and maritime dominance was rivalled by manufacturing wealth from south Lancashire, the West Riding of Yorkshire and elsewhere.

Figure 7.2   Highfield Cottage, near the foot of Coniston Water
Note: Typical of many smaller villas erected between 1815 and 1835.
Source: © The author.

Most villa owners, one supposes, had made the tour of the Lakes before deciding to build, and this was also true for many who, more cautiously, opted to rent. In some cases this can be demonstrated. Both William Wilberforce and William Gell left behind manuscript journals of their tours, made while undergraduates at Cambridge, Wilberforce in 1779 and Gell in 1797.[22] But while Wilberforce contented himself with returning in 1787–8 to rent Rayrigg, a minor gentry seat on Windermere, extended shortly afterwards to improve its eligibility as a villa, Gell built his own diminutive holiday retreat, known simply as 'Cottage, Grasmere' (1799), its scale probably dictated by the limited means of a young man not yet come into his inheritance.

Some villa builders were not tourists at all, but indigenous landowners who shared a similar appreciation of their native landscape and built either for themselves or speculatively, for sale or rent. One such was John Benson, Wordsworth's landlord at Dove Cottage, Grasmere, who before 1788 built and occupied Dove Nest on a prominent hillock above the increasingly celebrated Low Wood Inn on Windermere.[23] Benson died in 1808, and thereafter Dove Nest was occupied by a succession of tenants, most notably the poet Mrs Felicia Hemans in 1830. Benson had also owned a farmhouse in Grasmere known as Tail End. In 1809 this was let to Francis Duckinfield Astley, a colliery owner of Dukinfield, Manchester, who promptly added a new dining room to render it serviceable as a villa.[24]

Established villa-owners might also capitalise on the potential of their estates by building anew. A second villa could accommodate the overspill when large numbers of guests were invited, or it could enable guests to retain their privacy, and perhaps the use of their own servants, while visiting. John Marshall, the wealthy Leeds flax-spinner who built Hallsteads on Ullswater about 1815, used another villa known as Old Church to accommodate his wife's family. Thomas Dawson, who in 1834 bought Allan Bank, Grasmere, from John Gregory Crump (who had built it from 1805), built a cottage on the edge of his estate in 1837. It was occupied by the local curate, who later married one of Dawson's daughters, and the building, by now known as Glenthorne, was extended in 1867.[25] Many more owners extended existing houses, to the extent that it is rare to find a small villa from before 1850 that has not been considerably enlarged.

---

22   C.E. Wrangham (ed.), *Journey to the Lake District from Cambridge 1779: A diary written by William Wilberforce* (Stocksfield: Oriel Press, 1983); W. Gell, *A Tour in the Lakes 1797*, ed. W. Rollinson (Otley: Smith Settle, 2000).

23   See P. Crosthwaite's 1788 map of Windermere, in Hankinson, *Regatta Men*, unpaginated plates.

24   E. de Selincourt (ed.), *The Letters of William and Dorothy Wordsworth*, 2nd edn, rev. C.L. Shaver, M. Moorman and A.G. Hill, 8 vols (Oxford: Clarendon Press, 1967–93), vol. 2, pp. 375–6.

25   Deeds in National Trust files, supplemented by dates on the building.

All the villa colonies in the Lake District were rendered much more accessible by the building of railways and, to a lesser extent, the inauguration of connecting steamer services, particularly on Windermere and Ullswater. The railway reached Penrith (within striking distance of Ullswater) in 1846, Windermere in 1847 (with another branch reaching the southern tip of the lake in 1869, and connecting steamer services running to Ambleside), Coniston in 1859 and Keswick and Bassenthwaite Lake (on the trunk route between Penrith and Cockermouth) in 1864.[26] The western valleys (if one ignores the narrow-gauge Ravenglass & Eskdale Railway, built for mineral traffic) remained untouched, but the coastal route from Lancaster to Whitehaven and beyond, completed in 1857, brought them many hours closer to the main centres of population in England and Scotland. Villa owners were, on the face of it, among the principal beneficiaries of these schemes, but an analysis of those who subscribed to the committee opposing the Kendal & Windermere Railway (under the chairmanship of Professor Wilson of Elleray) reveals that a great many of them valued exclusivity more than convenience of access.[27] They may have deplored the prospect of the 'commuter villa' (in the 1870s Henry William Schneider of Belsfield, Bowness, famously commuted by steam yacht and private train to his ironworks in Barrow-in-Furness),[28] but the threat of factory excursion trains influenced them more decisively.

Tourism, Landscape and Society

The whole idea of the villa is inseparable from the possession of an area of garden ground. In the Lake District, with its rugged relief and mountainous horizons, 'possession' was easily extended to the appropriation of views well beyond the proprietary estate. Nevertheless, all but the most constrained of Lake District villas enjoyed their own private garden and usually some more extensive area of designed landscape. The largest also amassed farm and cottage property, or moorland suitable for the rearing and shooting of game birds, and these too might be refashioned to varying degrees. The need to provision the household – reducing dependence on local markets, which could become depleted at the height of the tourist season – made a kitchen garden nearly essential, and it was sometimes supplemented by a separate orchard and home farm. 'Home farm' is too exalted a description of some of these, which amounted to little more than a cow-house with a loft above for hay storage, sometimes combined with the stable and coach

---

26   D. Joy, *The Lake Counties: the Regional History of the Railways of Great Britain, Vol 14*, 2nd edn (Newton Abbot: David St John Thomas, 1990), pp. 195–222.

27   Public notice dated 1844, reproduced in D. Smith, *The Kendal & Windermere Railway: The first railway to be built into the heart of the Lake District* (Pinner: Cumbrian Railways Association, 2002), p. 6.

28   A.G. Banks, *H.W. Schneider of Barrow and Bowness* (Kendal: Titus Wilson, 1984), pp. 89–100.

house required by every rural gentleman's residence. Large villa estates might also include a gate lodge, summer house or ornamental building. Every lakeside villa had one or more boathouses to satisfy a mixture of practical, recreational and sporting needs.[29]

The preponderance of villa owners in the opposition to the Kendal & Windermere Railway reveals something of the social correlative of this landscape by the mid-nineteenth century. The society into which the villa-builders had moved was not quite Wordsworth's 'perfect republic of shepherds and agriculturists', but the existing gentry estates were few in number and modest in scale, and the larger landed estates (of the Lowthers and Dukes of Norfolk for example) were centred outside the Lake District proper. Those villa owners who became permanent or periodic residents were therefore quickly assimilated by, and ultimately came to dominate, the local social order. Their villas, gardens, landscaped parks and, in some cases, much wider estates, joined together in the larger villa colonies to create a nearly seamless landscape, and formed the domiciliary basis for a new social order in the Lake District (Figure 7.3).[30] This in turn found expression in an upsurge in the building of churches, schools and a range of other institutions and amenities – the collective impact of which, on key parts of the Lake District landscape, was immense. There are thus strong grounds for regarding the villa as central to a novel kind of landscape, one which is essentially defined by tourism, and one which overlies and revolutionises the earlier vernacular landscape of agriculture and rural industries. In Britain there are few if any inland areas of comparable extent which are as firmly marked by the imprint of tourism as the villa landscapes of the English Lake District.

The evolution of a single villa estate will illustrate some of the forces active in determining the shape of the modern Lake District landscape. Monk Coniston Hall (the name alludes to the medieval estates of Furness Abbey) stands on rising ground at the head of Coniston Water – its accustomed name down to the early nineteenth century was Waterhead House. A yeoman farmhouse stood on the site by the mid-seventeenth century when it belonged to the Harrisons.[31] This was replaced around the middle of the eighteenth century, probably by Richard Harrison (1676–1761). Following Harrison's death the property passed to his daughter, who had married William (1728–69), son of Richard Ford (1697–1757), ironmaster at Newlands near Ulverston. William Ford became a shareholder in the

---

29   A. Menuge, 'Patriotic Pleasures: Boathouses and Boating in the English Lakes', in Geoff Brandwood (ed.), *Living, Leisure and Law: Eight Building Types in England 1800–1914* (Reading: Spire Books, 2010), pp. 53–72.

30   For a study of the evolution of one villa colony, see O.M. Westall, 'The Retreat to Arcadia: Windermere as a Select Residential Resort in the Late-nineteenth Century', in idem (ed.), *Windermere in the Nineteenth Century*, rev. edn, (Lancaster: Centre for North-West Regional Studies, 1991), pp. 34–48.

31   I am indebted for the biographical details in this paragraph and subsequently to unpublished research by Janet Martin.

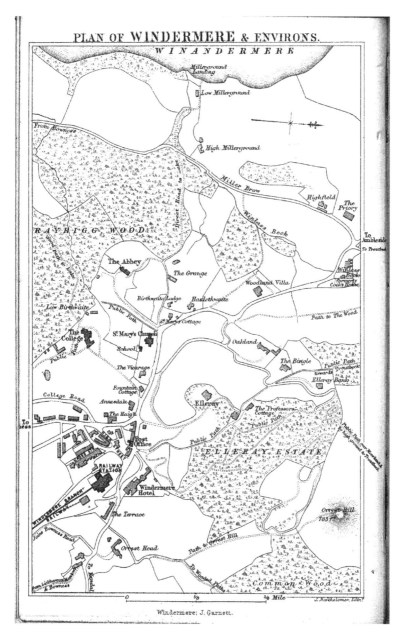

Figure 7.3 **Map of the nascent town of Windermere, published in 1871, vividly conveying the rapid growth of a wealthy and spacious villa colony close to the terminus of the Kendal & Windermere Railway**

Source: © Martineau 1871 (The author's collection).

firm, as did Michael Knott of Rydal and James Backhouse of Finsthwaite, who brought valuable coppice woods into the partnership at a time when the expansion of the iron industry was putting pressure on local charcoal production. After the early deaths of William Ford and his wife the Waterhead Estate descended by marriage to George, son of Michael Knott. It is with George Knott that we first hear of embellishments to the landscape,[32] but he was short-lived as well, and his son, also George, inherited while still a minor. The iron business was managed by trustees, under the name of Knott, Ainslie & Co., and in 1785 Waterhead House was let to a Londoner, Timothy Parker.

Neither the Harrisons nor the Fords lived the villa lifestyle. The comfortable, up-to-date but plain new house which replaced the earlier farmhouse had two storeys and attics, a three-bay main front and a subordinate kitchen wing to one side. But it was emphatically not a villa, showing all too clearly (as Wordsworth put it around 1809)

> a busy propensity ... preferring a view across the road & a blank hill side, with a patch [of] northern sky, to a noble prospect of the Lake stretching southward & which, tho' hidden from the house, may be seen from an alcove in the pleasure ground behind, at a short distance from the door ....[33]

What are the origins of this alcove? Coniston Water was not highly rated by early tourists, though many passed it en route for Hawkshead or Ambleside after approaching via the oversands route across Morecambe Bay. It is notable that it was not included among the first set of lake maps issued by the Keswick museum keeper Peter Crosthwaite in 1783, finding a place in his series only in 1788. Writers found its straightness less picturesque than the crooked and indented forms of Windermere, Ullswater and Derwent Water, and were perhaps also put off by the considerable copper-mining and slate-quarrying activity around its upper end. These factors were probably responsible for delaying the development of a villa colony here until the early years of the nineteenth century. There was advice to be had, nevertheless, on the best vantage points for viewing the lake and its adjacent mountains. West, in 1778, proposed four 'stations', three on the eastern shore and one on the surface of the lake, but others thought differently.[34] As early

---

32   T.W. Thompson, *Wordsworth's Hawkshead*, ed. R. Woof (Oxford: Oxford University Press, 1970), p. 27.

33   W.J.B. Owen and J.W. Smyser (eds), *The Prose Works of William Wordsworth*, 3 vols (Oxford: Oxford University Press, 1974), vol. 2, pp. 322–3.

34   Some travellers, like William Gell in 1797 and Richard Colt Hoare in 1800, omitted Coniston Water altogether from their itineraries (W. Gell, *A Tour in the Lakes 1797*, ed. William Rollinson (Otley: Smith Settle, 2000); M.W. Thompson (ed.), *The Journeys of Sir Richard Colt Hoare through Wales and England 1793–1810* (Gloucester: Alan Sutton, 1983), pp. 129–38).

as 1786 Yates showed a small building immediately east of Waterhead House,[35] and it is clear that this must be the 'alcove' (in reality a two-storey summer house) referred to by Wordsworth (Figure 7.4) It was probably the work of George Knott, forming part of wider landscape improvements which were subsequently enjoyed by Timothy Parker and painted by John Warwick Smith.[36] Crosthwaite made the summer house one of his three alternative Coniston Water 'stations'.

Figure 7.4   The summer house in the grounds of Monk Coniston Hall
Note: Visitors took the view from the upper room, which was entered from the artificially raised ground level to the right; the lower room incorporated a fireplace where refreshments could be heated.
Source: © The author.

---

35   W. Yates, *The County Palatine of Lancashire, surveyed by William Yates* (1786).

36   One of the 100 watercolours painted by Smith for John Christian Curwen of Belle Isle is titled 'From Mr Parker's Gardens, Coniston Lake' (*c.*1790), and depicts a terrace, open to the south and west and adorned with a sundial (see C. Powell and S. Hebron, *Savage Grandeur and Noblest Thoughts: Discovering the Lake District 1750–1820* (Grasmere: Wordsworth Trust, 2010), pp. 68–9).

George Knott was succeeded by his brother Michael in 1806. Michael Knott took a course familiar to students of British business history, selling his interest in the firm in 1812 and setting up as a leisured gentleman on the proceeds. He was clearly schooled in the appreciation of landscape, and shortly after he retired from business he set about remodelling Waterhead House. Knott retained the eighteenth-century house, adapting it to form a combined service wing and guest wing, and linking it to reception rooms and principal bedrooms in a large new block executed in the stiff, symmetrical Gothic that had come down from the eighteenth century. The reticence of the earlier house was exploited by retaining its approach road, which effected a characteristic deferral of gratification, so that the *coup de théâtre* burst on the visitor only once in one of the principal reception rooms. All three reception rooms looked out along a newly smoothed *clairvoyée*, beyond which stretched the whole length of Coniston Water. The prospect was flanked on one side by a crag, but on the other the view was left at least partly open, to disclose the mountains known as Old Man, Kitty Crag and Wetherlam, and more particularly the waterfall of White Gill, which descends spectacularly into Yewdale when in spate.

WATERHEAD HOUSE.
The residence of Michael Knott Esq.

**Figure 7.5   Stirzaker's portrayal of Monk Coniston Hall emphasised the
view (now largely obscured) towards the waterfall of White Gill**
Source: *Lonsdale Magazine*, May 1822 © Wordsworth Trust.

Michael Knott's building campaign was prolonged. Much had apparently been accomplished by September 1817, when Revd Francis Witts, a Gloucestershire tourist, noted a 'boat house ... in a bad taste of mock Gothic',[37] but work was still in progress on the house itself when William Green was preparing his 1819 guidebook.[38] It appears to have been complete by May 1822, when a fulsome description and an aquatint view by Richard Stirzaker (Figure 7.5), highlighting the view of the waterfall, were published in *The Lonsdale Magazine*.[39]

Knott racked up substantial debts through his building and landscaping works, and following his death in 1834 the estate, totalling 668 acres 2 roods and 15 perches, was offered for sale. Wordsworth alerted his friend John Marshall of Hallsteads (who was in the process of settling all his sons on Lake District estates) to the impending sale, noting that 'The improvements which Mr. K. has lately made there, are very great – and it is one of the most elegant residences in the Lake district'.[40] The purchaser was James Garth Marshall (1802–1873), and it was shortly afterwards that the house, in keeping with its newly Gothic appearance, began to be referred to as Monk Coniston Hall. James Garth Marshall made only a modest addition to the house, but he devoted substantial sums to creating the enchanting Tarn Hows landscape of woodland and lake, and he augmented the estate by the purchase, in 1839, of a neighbouring villa called Tent Lodge. Built around 1815 for Captain Smith, its name commemorating the tent in which his talented and erudite daughter Elizabeth had died of consumption in 1806, Tent Lodge brought with it a length of lake frontage, a boathouse, and an up-to-date house suitable for genteel visitors.

Marshall's business interests restricted his visits to Monk Coniston, but when there he dispensed hospitality prodigally if not always to his guests' entire satisfaction. The poet Tennyson spent his honeymoon at Tent Lodge in 1850, where he had 'but rough accommodations' – though not so rough as to prevent him returning for a six-week holiday in 1857.[41] Thomas Carlyle was less fortunate. He found his bedroom at Monk Coniston in 1850 (almost certainly in the converted old house) 'exposed to nurseries of children, tumults of flunkies & housemaids: I almost lost my sleep entirely that night'. He was found a quieter bedroom but felt that he had been invited principally for his reputation, and left after only two nights.[42] Other notable guests who experienced the Lakes through

37 A. Sutton (ed.), *The Complete Diary of a Cotswold Parson: The Diaries of the Revd Francis Edward Witts 1783–1854*, 10 vols (Chalford: Amberley, 2008), vol. 2, p. 234.

38 W. Green, *The Tourist's New Guide*, 2 vols (Kendal: R. Lough & Co., 1819), vol. 1, p. 119.

39 Anon., 'Beauties of the North: Waterhead House', *The Lonsdale Magazine and Kendal Repository* 3, 29 (31 May 1822), pp. 161–2 and plate facing p. 161.

40 M.E. Burton (ed.), *The Letters of Mary Wordsworth 1800–1855* (Oxford: Clarendon Press, 1958), p. 136.

41 C.Y. Lang and E.F. Shannon Jr (eds), *The Letters of Alfred Lord Tennyson*, 3 vols (Oxford: Clarendon Press, 1982–90), vol. 1, p. 332, and vol. 2, pp. 178 & 563–4.

42 C. de L. Ryals, K.J. Fielding et al. (eds), *The Collected Letters of Thomas and Jane Welsh Carlyle* (Durham and London: Duke University Press, 1970ff), vol. 25, pp. 232–46.

Marshall's hospitality included the poets Aubrey de Vere, Coventry Patmore and Matthew Arnold, the artists Thomas Woolner and Edward Lear, the diarist William Allingham, the astronomer Sir John Herschel and a young photographer named Charles Dodgson (Lewis Carroll). The pleasures of the estate included trips in jaunting cars, rowing on the lake and game-shooting.

Marshall's son Victor, who succeeded to the estate on the death of his mother in 1875, endured the collapse of the family firm in 1886, but twenty years later he still owned 4,179 acres, including the villas and smaller residences of Thwaite House, Thwaite Cottage, Tent Lodge and Tent Cottage, the Waterhead Hotel, a home farm, a dozen farmsteads, a series of slate quarries, an estate sawmill and numerous cottages.[43] On Victor's death in 1928 his son sold the bulk of the estate to Beatrix Potter, whose love of the Lake District had been nurtured during long childhood holidays at Wray Castle, Lingholm, Holehird and Lakefield – a list which includes some of the most substantial villas in the Lakes.[44]

Changing Occupants and Uses

'Retired merchants and professional men fall in love with the region, buy or build a house, are in a transport with what they have done, and, after a time, go away. In five or six years, six houses of friends or acquaintance of mine became inhabited by strangers'.[45] Thus Harriet Martineau, who moved to Ambleside in 1845, characterised the gap between intention and fulfilment among villa-builders. Most owners occupied their villas for only a few weeks of the year. Some turned up still less frequently, finding, once the initial rush of enthusiasm had subsided, that the arduous travel involved was a deterrent, or simply that the ties of society, kinship or business were more powerful than the allure of relatively remote and esoteric pleasures. Their properties joined those built speculatively for lease or rent. There were thus plenty of opportunities for those who could not afford, or did not choose, to own a villa to take one for a holiday or season, and sometimes to move into one, furnished or unfurnished, on a longer lease.

In attracting short-term occupiers villas competed with a range of other providers, including hotels, inns and lodging houses, as well as more informal arrangements where a household took in friends or paying guests. Harriet Martineau herself began by taking lodgings at Ambleside before building her compact villa, The Knoll, in 1845–1846, but

---

43   1908 Sale Particulars, CRO(K), WD/B/35/SP252.

44   L. Linder (ed.), *The Journal of Beatrix Potter from 1881 to 1897* (London and New York: F. Warne, 1966), pp. 19–23, 147–51, 382–95 and 416–23.

45   H. Martineau, *Autobiography*, 3rd edn, 3 vols (London: Smith, Elder & Co, 1877), vol. 2, p. 221.

soon found that I must pay a serious tax for living in my paradise: I must, like many of my neighbours, go away in 'the tourist season.' My practice has since been to let my house for the months of July, August and September, – or for the two latter at least, and go to the sea, or some country place where I could be quiet.[46]

Of the different types of accommodation available to tourists, hotels and inns were invariably the most expensive, though for those whose time in the Lakes was limited, and who were therefore obliged to shift their abode in order to see the sights, they were a practical necessity. Villas were a cost-effective alternative wherever a stay of some weeks or months was contemplated, and the demand for them in the peak season was brisk. Those who secured one enjoyed a number of advantages. They might have ready-made introductions to local society and could entertain company if they chose. They were also equipped for days of rest as well as days of exertion and, if they could afford one of the better situated houses, they could enjoy enviable private grounds.

With the arrival of the railway in the 1840s the demand for accommodation intensified and a number of villas were adapted to serve as hotels. The Hollens, Grasmere, had become the Hollins and Lowther Hotel by 1848 under the proprietorship of Edward Brown, previously an innkeeper in Penrith. The timing of the change suggests the influence of the Kendal & Windermere Railway just a few miles distant. Before 1852 the Kendal architects, Webster and Thompson, had remodelled the modest cottage villa with its symmetrical single-storey wings, raising one to two storeys and the other to three, in the form of a prominent, asymmetrical Italianate tower.[47] The economic and social changes ushered in by the railway were felt particularly in Bowness. Here the oddly named Old England, a villa dating from shortly before 1819, was adapted to serve as a hotel in 1859 and rebuilt ten years later.[48] Nearby Belsfield, built in 1845 by Webster – with an original Italianate tower – for the Baroness de Sternberg, and later home to the industrialist Henry William Schneider, became the Belsfield Hotel in 1890. The Storrs Hall Hotel opened two years later, following the addition of a bulky range of hotel bedrooms.

After the First World War the role of the villa in the tourist economy of the Lake District began to change dramatically. The number of villas in private hands slowly dwindled, with the commercial and institutional sectors being the main beneficiaries. The end of the twentieth century saw something of a resurgence of private investment, but typically this resulted in the subdivision of villas, and often their ancillary ranges, to form apartments.

---

46   Ibid., p. 266.

47   A. Taylor, *The Websters of Kendal: A North-Western Architectural Dynasty*, ed. J. Martin, Cumberland & Westmorland Antiq. & Archaeol. Soc. Record Series 17 (Kendal, 2004), p. 162.

48   M. Hyde, N. Pevsner et al., *The Buildings of England: Cumbria* (New Haven and London: Yale University Press, 2010), pp. 168–9.

Conversions to hotel use continued after the Second World War. Sharrowbay (Figure 7.6), which was built on the southern shore of Ullswater in 1844, claims to be 'the original country house hotel'; it was certainly a brave venture, opening in the still-rationed Britain of 1949.[49] Today the 'boutique' hotel sector is much more extensive. By contrast some institutional owners acquired villas to provide low-cost holiday accommodation to the expanding tourist market, and especially to make the Lake District hospitable to young people. The Youth Hostel Association began a long association with the Lake District villa in or shortly before the Second World War, when it acquired Esthwaite Lodge, Hawkshead, going on to add Pocklington's Barrow House some years later. The Holiday Fellowship, an organisation set up in 1913 by T.A. Leonard to make 'open-air' holidays more accessible, was not far behind, leasing Monk Coniston Hall from the National Trust in 1945, and others followed: the Outward Bound Association acquired John Marshall's Hallsteads, and the similarly constituted Brathay Trust has occupied Brathay Hall since 1946. A number of villas are used by schools for similar purposes or by faith organisations (such as the Society of Friends at Glenthorne, Grasmere) providing a setting for more contemplative recreation.

Figure 7.6   Sharrowbay, now Sharrow Bay Country House Hotel, exemplifies the dilution of strictly proprietorial exclusivity in the Lake District

Source: © The author.

---

49   Sharrowbay Hotel brochure, 2010.

Conclusion

The Lake District villa owes its existence to the upsurge of interest in mountain and lake scenery during the second half of the eighteenth century, part of a wider transvaluation in which not only the picturesque taste but also a major strand of the Romantic Movement is rooted. It was fostered by growing wealth and leisure among a small but significant section of English society, and facilitated by improved communications, initially by road and sea, and from the 1840s by rail. The distribution of villas in a series of discrete clusters reflects a broad and surprisingly enduring consensus as to the relative landscape merits of different valleys, lakes and views, but it is modified by practical considerations of access and the availability of requisite services. As such, the distribution reinforces the popularity of certain favoured areas and locations, tending to the production of a dichotomous landscape in which a core area is simultaneously 'civilised' and 'consumed' by villa-building and the interlocking of polite landscapes, while the remainder (notably the rugged western valleys) continues untamed. Yet popular perceptions of the English Lake District remain stubbornly mired in an uncritical belief in its essentially 'natural' character. Closer attention to the villa and its landscape helps to 'civilise' this untutored perception, and to arrive at a better measure of the cultural processes shaping one of our most treasured national assets.

Chapter 8

The Origins and Development of Mountaineering and Rock Climbing Tourism in the Lake District, *c*.1800–1914

Jonathan Westaway

The Lake District has played a central role in the development of the sport of rock climbing in the British Isles, the late nineteenth century witnessing increasingly athletic and gymnastic approaches to the crags and fells. Climbing as we know it today emerged as a sub-cultural development, gradually establishing the legitimacy of an increasingly sporting approach to mountains, thereby changing the accepted notions of what it meant to be a mountaineer. Despite the accumulated Romantic and literary reputations of the Lake District landscape, mountaineering there was slow to emerge, the cultural significance of the Lake District landscape to mountaineers remaining for most of the nineteenth century, if not entirely negligible, then certainly subordinate. Why this should be so forms the basis of the first part of this chapter, focussing on the cultural significance of mountains in a European context and the pre-eminence of Alpine mountain landscapes as tourist destinations amongst the bourgeoisie. The elite tourist focus on 'being elsewhere'[1], on social distancing, exclusive locations and internalised notions of Romantic individualism and self-cultivation, meant that Alpine mountain landscapes emerged as the locus of the British middle classes' fascination with mountaineering in the middle of the nineteenth century. The second half of this chapter examines the emergence of the Lake District as a tourist destination for British mountaineers and rock climbers in the 1880s and 1890s as the regional middle classes sought to exploit the recreational hinterlands of the industrial cities of the north of England by adapting elite cultural approaches to mountain landscapes and applying them to regional upland landscapes. Part of a drive to express regional self-identity and exceptionalism, the cultural meaning of the Lake District landscape to mountaineers and climbers, always multi-layered, became enmeshed in new recreational paradigms, gradually differentiating itself from the Alpine mountaineering tradition that had gone before.

---

1   S. Baranowski and E. Furlough (eds.), *Being Elsewhere: Tourism, Consumer Culture and Identity in Modern Europe and North America* (Ann Arbor: University of Michigan Press, 2001).

Historians seeking to explain the origins of mountaineering and rock climbing in the British Isles have struggled to present a coherent account of how, but more importantly why, mountaineering as a leisure and sporting activity emerged in the middle of the nineteenth century. Historians of sport have found it particularly difficult to categorise, for although it contained 'sporting' elements and motivations, it reached its apotheosis in the middle of the nineteenth century as a leisure activity amongst upper-middle class British tourists whilst on vacation in the Alps. Unlike the dominant British sporting paradigm, the team-sports model, British mountaineers seemed reluctant to form clubs, or indeed any form of institutional voluntary association that would celebrate and facilitate recreational approaches to native mountains. It presented the peculiar spectacle of a major British sporting and recreational pastime that took place entirely on the Continent. The second half of the nineteenth century saw the slow transformation of mountaineering from a social field dominated by notions of romantic individualism and cultivated excursionism, where external aesthetic, literary, cultural and scientific criteria valorised behaviour, to a social field where internal and sub-cultural criteria increasingly determined value.[2] The growth of elements of sporting culture, with an emphasis on rock climbing in the British Isles, was the result of widening class participation in the 1880s and 90s.[3] This was not always welcomed or encouraged by the Alpine mountaineering establishment.

We cannot begin to understand the emergence of rock climbing and mountaineering in the Lake District without an analysis of class. Alpine mountaineering was a form of cultural capital that required time, money, motivation and cultivation to acquire. To be a mountaineer in Britain for most of the nineteenth century meant to climb in the Alps, Europe or the Greater Ranges. The Alpine Club, founded in 1857 by the pioneers of British mountaineering, was based in London. Its members were mainly upper-middle class professionals, and 'more likely to be Liberal Dissenters than Tory Anglicans'.[4] The club resembled other elite gentlemen's clubs, providing space and opportunity for expressions of shared interests, offering select conviviality, fellowship and exclusivity. The goal of the Alpine Club was not to establish a national mountaineering movement with local representation and regional affiliates with local mountain affinities, or

---

2   O. Hoibian, 'Sociogenesis of a Social Field: the cultural world of mountaineering in France from 1870 to 1930', *International Review for the Sociology of Sport* 41 (2006), pp. 340–1.

3   J.H. Westaway, 'The German Community in Manchester, middle-class culture and the development of mountaineering in Britain, *c*.1850–1914', *English Historical Review* 124 (2008), pp. 592–95.

4   P.H. Hansen, 'Albert Smith, the Alpine Club, and the invention of mountaineering in mid-Victorian Britain', *Journal of British Studies* 34 (1995), pp. 310–11; P.H. Hansen, *British Mountaineering, 1850–1914*, Harvard University, PhD thesis, chapter 2, 'The Alpine Club and Alpine Exploration', pp.107–61; D. Robertson, 'Mid-Victorians amongst the Alps', in U.C. Knoepflmacher and G.B. Tennyson (eds), *Nature and the Victorian Imagination* (Berkeley: University of California Press, 1977), p. 120.

to encourage mass tourism, healthy lifestyles or lower-middle class and working class participation. It existed to celebrate its members' achievements and maintain its exclusivity. Membership required Alpine climbing experience 'or evidence of literary or artistic accomplishments related to mountains'.[5] Alpine mountaineering had high barriers to entry in terms of the costs of an Alpine holiday and this exclusivity was cultivated and reinforced by devotees of mountaineering when they met in London. What constituted the correct approach to mountaineering and mountain landscapes was maintained by cultural arbiters in the Alpine Club, prospective members having to negotiate the usual clubland paraphernalia of proposal, election and exclusion.[6]

> A de facto social qualification also contributed to the genteel identity of the Alpine Club as 'a club for gentlemen who also climb.' These policies ensured that the social profile of the Alpine Club remained somewhat higher, and the number of its members much lower, than the mountaineering clubs founded throughout Europe in the 1860s and 1870s.[7]

The focus on Alpine landscapes as the premier tourist destination for British mountaineers coloured impressions of upland Britain. For most of the nineteenth century the mountains of the Lake District were understood in comparison to the Alps. This paradox was a commonplace of aestheticised approaches to mountain landscapes. What was true of the picturesque tourist in the eighteenth century was true of the mountaineering tourist in the nineteenth century: 'Educated awareness of what constitutes an ideal landscape' meant that 'the tourist travelling through the Lakes and North Wales will loudly acclaim the *native* beauties of British landscape by invoking idealised *foreign* models'.[8] As the hills of the Lake District offered no Alpine-scale challenges, or unclimbed summits, they received relatively little systematic attention from mountaineers affiliated with the Alpine Club in the first two decades of the club's existence, and a good deal of condescension. Ironically perhaps, it was the voluminous popular literature produced by Alpine mountaineers, where Alpine landscapes were idealised, aestheticised and dramatised, that provided many people's first encounter with the Alps and with mountaineers' cultural approaches to mountains. 'Landscapes', as Simon Schama has reminded us, 'are culture before they are nature; constructs of the imagination'.[9] For the majority of tourists and mountaineers, the mountains and fells of the Lake District were experienced as ideas before they were ever experienced in reality, ideas formulated and articulated by a particular class at a

---

5   Hansen, 'Albert Smith', pp. 309–10.

6   A. Lunn, *Mountain Jubilee* (London: Eyre & Spottiswoode, 1943), pp. 205–6.

7   Hansen, 'Albert Smith', p. 310.

8   M. Andrews, *The Search for the Picturesque: Landscape Aesthetics and Tourism in Britain, 1760–1800* (Stanford, California: Stanford University Press, 1989), p. 3.

9   S. Schama, *Landscape and Memory* (London: Harper Collins, 1995), p. 61.

particular time. We cannot understand the cultural meaning of the mountains of the Lake District to mountaineers and rock climbers without a wider understanding of cultural approaches to mountains amongst the European bourgeoisie that had cohered by the middle decades of the nineteenth century.

Undoubtedly tectonic upheavals in human intellectual perception across Europe in the eighteenth century shaped attitudes to mountain landscapes. The Enlightenment reversed the polarities of aesthetic perceptions, theories of the sublime inserting themselves between nature and the human soul, expressed as a desire for a new 'sensation of wonder mixed with fear, a pleasurable encounter with forbidding landscape or the darker passions'.[10] Natural philosophers, natural historians and scientists began to seek evidence for God's handiwork in the natural world. In seeking evidence of design they stumbled upon facts that altered their notions of chronology and man's place in the universe. Geologists and glaciologists found in the mountains evidence of deep-time, recalibrating our notions of the scale of evolutionary forces, relegating notions of God's agency to an ever more remote singularity. With the 1779 publication of *Voyage dans les Alpes* by de Saussure, Professor of Natural Philosophy at the Geneva Academy, the notion of climbing a mountain for scientific purposes was firmly established.[11] As notions of God's agency and immanency began their long recessional, Enlightenment *philosophes* celebrated the perfectibility of mankind, of man in a state of nature, exemplifying the nobility of peasant communities in mountain regions, of mountain landscapes as the very 'Seat of Virtue'.[12] Rousseau's *Nouvelle Héloïse* (1760) 'was chiefly responsible for focusing changing perspectives on the Alps'.[13] Mountain worship was a feature of the German *Sturm und Drang* movement in literature, where sublimity and subjective experience formed a literary topos as jagged and extreme as the mountain landscapes in which they were set.[14] German Idealist thought found in the struggle to climb a mountain and attain a summit to be the perfect metaphor for the dialectical method and the teleological drive towards the Ideal, exemplified

---

10   F. Spufford, *I May Be Some Time: Ice and the English Imagination* (London: Faber & Faber, 1996), p. 18.

11   J. Ring, *How the English Made the Alps* (London: John Murray, 2000), p. 19; C. Bigg, D. Aubin and P. Felsch, 'Introduction: the laboratory of nature – science in the mountains', *Science in Context* 23 (2009), pp. 311–21.

12   Schama, *Landscape and Memory*, pp. 478–90; G. Rudaz., 'Stewards of the Mountains: The Poetics and Politics of Local Knowledge in the Valasian Alps', in D. Cosgrove and V. della Dora (eds), *High Places: Cultural Geographies of Mountains, Ice and Science* (London: I.B. Tauris, 2009), p. 153.

13   Ring, *How the English Made the Alps*, p. 17; Schama, *Landscape and Memory*, pp. 478–90.

14   P. Giacomoni., 'Mountain Landscapes and the Aesthetics of the Sublime in Romantic Narration', in G.E.P. Gillespie, M. Engel and B. Dieterlie (eds), *Romantic Prose Fiction - Comparative Histories of Literatures in European Languages* XXIII (Amsterdam/ Philadelphia: John Benjamins Publishing, 2008), pp. 111–12.

in Goethe's *Faust*. This reverence for mountain landscapes and their symbolic importance was further reinforced in German culture by romantic nationalism.[15] Since the publication of Tieck's novel *The Wandering of Franz Sternbald* (1798) 'the walking and sketching tour became an essential part of young Germans' self-discovery, and their interest focussed increasingly on the beauties of their homeland'.[16] Tieck, echoing Schelling's *Naturephilosophie* in his novels, did much to romanticise the 'wild call of the mountains'.[17] It was further expressed by the Turner movement's devotion to physical renewal in an outdoor context as a way of rebuilding the individual and society.[18]

These currents in European thought profoundly influenced English Romanticism and the Lakeland poets' perception of the Lake District landscape, none more so than Coleridge. Before Coleridge had even set eyes on the Lake District he had attended the University of Göttingen and climbed the Brocken in the Hartz mountains.[19] A significant translator and interpreter of German Idealist thought and literature, he had also imbibed the German cultural reverence for mountain landscapes. In August 1802 he set off on a nine day tour from Keswick to St. Bees on the Cumberland coast and back over the central fells of the Lake District, maintaining a letter journal which his biographer describes as forming 'the first literary description of the peculiarly English sport of fell-walking. Coleridge was in effect inventing a new kind of Romantic tourism, abandoning the coach and the high-road for the hill, the flask and the knapsack'.[20] He also largely abandoned guide books which 'tended to provide travellers with fixed itineraries around a series of principal sights'.[21] He took a sketch map in his pocket book, 'a chart of the mountainous, craggy countryside over which he planned to wander'.[22] On 5 August on the summit of Scafell Coleridge penned a letter describing what he saw and how he felt:

---

15 W. Graeber, 'Nature and Landscape between Exoticism and National areas of Imagination', in Gillespie, Engel and Dieterlie (eds) *Romantic Prose Fiction*, pp. 94–6.

16 A. Griffiths and F. Carey, *German Printmaking in the Age of Goethe* (London: British Museum Press, 1994), p. 216.

17 Giacomoni, 'Mountain Landscapes', p. 115.

18 I. Weiler, 'The Living Legacy: Classical Sport and Nineteenth-Century Middle-Class Commentators of the German-Speaking Nations', in J.A. Mangan, (ed.), *The European Sports History Review, iv: Reformers, Sport, Modernizers: Middle-Class Revolutionaries* (London: Frank Cass, 2002), p. 10; C. Eisenberg, 'Charismatic National leader: Turnvater Jahn', in R. Holt, J.A. Mangan and P. Lanfranchi (eds), *European Heroes: Myth, Identity, Sport* (London: Frank Cass, 1996), pp. 14–27.

19 R. Holmes, *Coleridge: Early Visions* (London: Harper Collins, 1998), pp. 229–31.

20 Ibid., p. 328.

21 R. Hewitt, *Map of a Nation: A Biography of the Ordnance Survey* (London: Granta, 2010), p. 209.

22 Ibid., p. 210.

O my God! What enourmous Mountains these are close to me … the Clouds are
hast'ning hither from the Sea – and the whole air seaward has a lurid Look – and
we shall certainly have Thunder.[23]

He then began to descend Broad Stand, today considered to be a Moderate
grade rock climb. Lowering himself over edges he soon found he could not
go back. His own account of the descent 'shows Coleridge at his finest pitch:
a comic hero beset by tragic visions, spiritual, intelligent, and supremely self-
aware of his own psychological drama', the exertion 'making his muscles tremble
with exhaustion and vertigo', his mind exulting, producing an 'effect of almost
religious intensity'.[24] Coleridge's descent of Broad Stand to Mickledore has been
appropriated as foundational by British mountaineering historians searching for
antecedents.[25] Whilst Coleridge, the 'Metaphysical Mountaineer'[26] was also an
inadvertent one, this climbing interlude and the wider mountain tour of which it
was part represents a significant cultural shift in attitudes to mountains, unleashing
the intellectual forces that would shape mountaineering in the nineteenth century.

Since at least the 1770s the Lakeland scenery of Lancashire, Cumberland
and Westmorland had been 'a serious challenge to the aesthetic supremacy of
the European Grand Tour',[27] the tour to the Lakes a necessary component of
elite tourist itineraries. The Picturesque tourist appropriation of landscape was,
however, essentially passive and contemplative, effected by means of seeing and
hearing. Whilst Gray's *Journal of the Lakes* (1769) had transposed the sublime's
'vocabulary of "horrid beauty"' from the Alps to the Lakes, establishing the Lake
District as the definitively sublime English landscape',[28] it took the Romantic
sensibility to accelerate beyond this, touching, feeling and experiencing landscape
in ways that were often visceral, ludic and pantheistic. Above all, Romanticism
stressed the primacy of the individual. It valorised the subjective experience of
landscape and mythologised what we might call a wilder tourism, at the heart
of which was the notion of the solitary wanderer, the romantic individual.
Romanticised individualism and aestheticised approaches to landscape was a
feature of pan-European bourgeois notions of *Bildung*

> which stressed inner growth, the development of the whole personality and
> the elevation and transformation of the individual through contact with classic
> works of civilization and sublime natural landmarks. … Visual connoisseurship
> over "nature" – its study, its appreciation, its proper care – provided the educated

---

23   Holmes, *Coleridge*, p. 329.
24   Ibid., p. 330.
25   A. Hankinson, *A Century on the Crags: The Story of Rock Climbing in the Lake
District* (London: J. M. Dent & Sons, 1988), p. 14.
26   Holmes, *Coleridge*, chapter 13.
27   Andrews, *The Search for the Picturesque*, p. 153.
28   Schama, *Landscape and Memory*, p. 471.

middle class with an index of good breeding, a seemingly irrefutable ground
from which to lay claim to social and cultural leadership.[29]

Mountaineering, a largely upper-middle class pursuit of the professional classes,
was not immune from these class-based assumptions. Mountaineers, exemplifying
the educated bourgeoisie, conceived of themselves as travellers (as opposed to
tourists) satisfying wider aesthetic, cultural and scientific criteria. The end of the
Napoleonic Wars saw a boom in British tourists reacquainting themselves with
the Alps.[30] Romantic exiles like Shelley and Byron eulogised Swiss democracy,
Rousseauvian simplicity and the 'cold sublimity' of the Alps: 'To me', said
Lord Byron, 'High mountains are a feeling'.[31] Increasingly the Romantic desire
to experience and not just contemplate mountains was fused with a desire to
understand their origins. British scientists like Forbes and Tyndall popularised the
Alps with their writings.[32] As tourist infrastructure developed in the 1840s and as
Switzerland settled down after the *Sonderbundskrieg* of 1847 there was a huge
influx of British tourists in the Alps. The railways reached Basel and Geneva and
'Switzerland came to be hardly more than a day's journey from London and the
fare less than ten pounds'.[33] The 1850s and 1860s were to see a huge explosion
in British mountaineering in Alps. British mountaineering historians have
characterised this as the 'Golden Age' of mountaineering. Between Alfred Wills'
ascent of the Wetterhorn from Grindelwald in September 1854 and Whymper's
ascent (and subsequent disaster) on the Matterhorn in 1865, a huge number of
Alpine peaks were climbed for the first time. 'Of forty-three first ascents in 1864
and 1865, for example, only five had been made by continental climbers'.[34] It was
not just the fact of these achievements but their representation in print that made
these mountaineering endeavours celebrated and familiar, drawing further tourists
to the Alps in search of adventure. Wills' ascent of the Wetterhorn was not a first

---

29   T. Lekan, 'A "Noble Prospect": Tourism, *Heimat,* and Conservation on the Rhine,
1880–1914', *Journal of Modern History* 81 (2009), p. 844.

30   J. Buzard, *The Beaten Track: European Tourism, Literature, and the Ways to
"Culture", 1800–1918* (Oxford: Clarendon Press, 1998), p . 80: L. Withey, *Grand Tours
and Cook's Tours: A History of Leisure Travel, 1750 to 1915,* (New York: William Morrow
& Co, 1997), pp. 196–219.

31   P.J. Manning (ed.), *Lord Byron: Selected Poems* (London: Penguin, 2005), Childe
Harold's Pilgrimage: Canto III, pp. 415–55, at 436, 440.

32   F. Fleming, *Killing Dragons: The Conquest of the Alps* (London: Granta, 2000),
pp. 126–45, 177–91; T. Braham, *When the Alps Cast Their Spell: Mountaineers of the
Alpine Golden Age* (Glasgow: The In Pinn, 2004), pp. 50–80; C.E. Engel, *Mountaineering
in the Alps: a Historical Survey* (London: George Allen & Unwin, 1971), pp. 84–107;
M.S. Reidy, 'John Tyndall's Vertical Physics: from rock quarries to icy peaks', *Physics in
Perspective* 12 (2010), pp. 122–45.

33   N. Annan, *Leslie Stephen: The Godless Victorian* (London: Weidenfeld &
Nicholson, 1984), p. 90.

34   Fleming, *Killing Dragons,* p. 291

ascent but the fifth or sixth; 'it was not even the first ascent made by an amateur, for Agassiz had been up the mountain ten years earlier. But it was the first ascent to inspire a well-written narrative'.[35] The Alpine Club anthology *Peaks, Passes and Glaciers* was a great publishing success in 1859 and Whymper, capitalising on his notoriety after acres of newspaper coverage of the Matterhorn disaster, produced *Scrambles Amongst the Alps* (1871), deemed by many of his contemporaries to be the ultimate expression of Alpinism.

Historians seeking to understand this new British tourist *imperium* in the Alps and explain why people began to climb mountains for recreational purposes have outlined new historical forces at work. Dominant cultural discourses within British imperialism that stressed manliness and athleticism saw a growth of middle-class interest in sport and body cultivation. The British public schools promoted new chivalric codes that stressed stoicism and endurance, shaping perceptions of how to behave. From the peripheries of empire came tales of adventure and exploration that stimulated the desire to emulate and embody imperial vigour. If, as one commentator has noted, achievement was the 'essence of mountaineering' then 'climbing gives, to those who need it, the reassurance that they are men – men still capable of defeating the tyranny of life'.[36] In the minds of many tourists, the Alps became a second Pole, the one month holiday in the Alps an opportunity to enact a type of imperial manliness.[37] Thus the specificity of the British bourgeois experience added new strata of meaning to the experience of mountain landscapes. To visual, aesthetic and cultural connoisseurship over nature the upper-middle class mountaineering tourist added notions of exploration and conquest. The impulse to dominate the Alps, evidenced by the first ascents of unclimbed peaks, suggests not only a desire to overcome nature but to establish primacy, a form of territoriality often at odds with the nationalist objectives of Alpine nations themselves.

But the bourgeoisie was by its nature cosmopolitan, subject to wider intellectual forces beyond the imperial domain. The professional and intellectual classes, struggling with the crises of faith, the question of what it means to be human and to lead the good life, found in mountains the space for inner exploration, spiritual solace and teleological hopes. The scale of Alpine mountains inspired awe, the evidence of geological time became 'akin to time travel: a way to access the perspectives of the planet, if not the universe'.[38] This Copernican moment

---

35   Engel, *Mountaineering in the Alps*, p. 93.

36   Annan, *Leslie Stephen*, p. 92.

37   This discourse of conquest is perhaps most clearly articulated in the work of Peter H. Hansen. See, for instance, P.H. Hansen, 'Modern Mountains: The Performative Consciousness of Modernity in Britain, 1870–1940', in M.J. Daunton and R. Bernhard (eds), *Meaning and Modernity: Britain in the Age of Imperialism and World Wars* (Oxford: Berg, 2001) p. 187. D. Robertson, 'Mid-Victorians amongst the Alps', p. 133 was of the opinion that the language of conquest was rarely employed by writers in the pages of the *Alpine Journal*.

38   Schama, *Landscape and Memory*, p. 488.

opened up new prospects to the Victorian upper-middle class mountaineer: 'the endless space of our interior self'.[39] Alpine mysticism re-enchanted the inner lives of Victorian doubters, sceptics and rationalists, recovering in mountain landscapes 'that "soul of loveliness" which the universe appeared to have lost'.[40] Leslie Stephen, author of *An Agnostic's Apology*, ruminated that if he were to invent a new idolatry 'I should prostrate myself, not before beast, or ocean, or sun, but before one of those gigantic masses, to which, in spite of all reason, it is impossible not to attribute some shadowy personality'.[41] For Victorian intellectuals, the Alps took on aspects of the deity. The valetudinarian John Addington Symonds declared: 'As I am prostrated and rendered vacant by scepticism, the Alps are my God. I can rest there and feel, if not God, at least greatness – greatness prior, and posterior to man in time, beyond his thoughts, not of his creation, independent, palpable, immovable, proved'.[42] Mountaineering, as Noel Annan has noted, had a peculiar hold over intellectuals:

> Mountaineering makes it possible for the intellectual to experience things which would otherwise be impossible: danger, intense comradeship, manliness, physical pain in pursuit of tangible objective, and a sensation of being at one with Nature. Psychologically, the intellectual is always conscious of his own isolation. He is not at one with the human race.[43]

These class-based attitudes coloured mountaineers' notions of how to see, appreciate and experience mountains. The upper-middle class professionals of the Alpine Club considered themselves a 'natural aristocracy ... that turned its back on the industrial world of gutta-percha shoddiness'.[44]

To a self-consciously intellectual aristocracy,[45] Alpine mysticism offered expiatory '*Selbst-tödtung*, a total surrender to larger energies ... here the end is

---

39   Ibid., p. 489.

40   Lunn, *Mountain Jubilee*, p. 73.

41   Ibid.

42   B. Schultz, *Henry Sidgwick: Eye of the Universe* (Cambridge: Cambridge University Press, 2004), p. 403.

43   Annan, *Leslie Stephen: The Godless Victorian*, p. 91.

44   Schama, *Landscape and Memory*, pp. 502–3.

45   Robertson, 'Mid-Victorians amongst the Alps', pp. 120–22 makes the point that early mountaineers were from overwhelmingly intellectual professions, drawing on Noel Annan's seminal essay 'The Intellectual Aristocracy', first published in J.H. Plumb (ed.), *Studies in Social History: A Tribute to G. M. Trevelyan* (London: Longmans, 1955), pp. 243–87 and since republished as an appendix to N. Annan, *The Dons: Mentors, Eccentrics and Geniuses* (London: Harper Collins, 1999), pp. 304–41. C.E. Matthews, speaking as President of the newly founded Climbers' Club in 1898 and referring to the present composition of the Alpine Club noted 'it comprises within its ranks some of the best of the intellectual aristocracy of this country'. See G.E. Bryant, 'The Formation of the Climbers' Club', *The Climbers' Club Journal* 1, 1 (August 1898), p. 6. See also Hansen's analysis that

not connection with humanity but a transcendence of it'.[46] This resonated with those seeking distance from society.[47] Stefan Collini has noted what he terms a 'Muscular Liberalism' as a defining characteristic of mid-Victorian Liberal intellectual elites, exhibiting and valuing stoicism and strenuousness, alive to the 'ennobling compulsion of struggle',[48] where a largely secular ethos of manliness was synonymous with individualism and 'weakness of will ... could be walked or climbed out of the system'.[49]

Given this emphasis on individualism, exclusivity and intellectualism, bourgeois-romantic anti-tourism was never far below the surface in Alpine mountaineering circles, critical of the spread of mass tourism to the lower-middle classes and the working classes, and always ready to assert aesthetic and cultural superiority.[50] Alpine mountaineers' self-definition was constructed in contradistinction to mass tourism and class-based snobberies were commonly expressed.[51] In Britain, social stratification meant that mountaineering remained the preserve of an elite. Liberal individualism was hostile to the notion of mass participation. Elite tourist destinations emerged in the Alps, trying to keep one step ahead of Cook's tourists. The British *rentier* class de-camped to Switzerland in search of a healthier lifestyle, cheap accommodation, mountain scenery and cures for tuberculosis. Winter sports emerged which were socially selective from their inception. The aspirations and affinities of the bourgeoisie, often trans-national in composition, meant that 'being elsewhere' was always an option. The 'compensatory domains' carved out by elite tourists represented jealously guarded freedom for the privileged few:

'the Alpine Club appears to be more representative of the larger and more diverse group of "gentlemanly capitalists."': Hansen, 'Albert Smith', p. 311.

46   G. Levine, 'High and Low: Ruskin and the Novelists', in Knoepflmacher and Tennyson, *Nature and the Victorian Imagination*, p. 141. 'Prostration before sublimity' led Levine to label Alpine mysticism 'a mistake, a false tourism'.

47   Y.-F. Tuan, Review of Cosgrove and della Dora, (eds), *High Places*, in *Progress in Human Geography* 34 (2010), p. 403.

48   Schama., *Landscape and Memory*, p. 503.

49   S. Collini, *Public Moralists: Political Thought and Intellectual Life in Britain 1850–1930* (Oxford: Clarendon Press, 1991), pp. 113, 170–96. See also Leonard Woolf's comment that G.M. Treveleyan, a great lover of strenuous walking in the hills, was a 'Muscular agnostic': D. Cannadine, *G.M. Treveleyan: A Life in History* (London: Harper Collins, 1992), p. 145.

50   B. Debarbieux, 'Construits identitaires et imaginaries de la territorialité: variations autour de la figure du "montagnard"', preprint on author's webpage http://www.unige.ch/ses/geo/collaborateurs/publicationsBD.html, originally published as 'Le montagnard : imaginaires de la territorialité et invention d'un type humain', *Annales de Géographie* 660 (2008), pp. 90–115.

51   Annan, *Leslie Stephen*, p. 235; Bryant, 'The Formation of the Climbers' Club', p. 8.

Temporarily removing one from domestic society, the tour abroad presents an image in high relief of culture's potential function in modern industrial democracies: the cultural is conceived of as "outside" ordinary social life, comprising a compensatory domain of autonomy and creativity to which utilitarian capitalist social arrangements pay no heed. Travel, like culture, offers an imaginative freedom not as a rule available in modern social life.[52]

For the upper middle class British mountaineer, the Alpine mountaineering holiday offered freedom, a liminal space outside of economy and society, where liberal individualism and personal liberty were maximised, small group fellowship was forged and exclusivity maintained. The persistence of notions of romantic individualism and *bildung* meant that those elements of the British middle classes that could have demonstrated social leadership within the wider outdoor movement were reluctant to participate in voluntary associations that would widen participation. In Britain this delayed the development of mountaineering and climbing clubs with cross-class appeal. In the European Alpine nations in the 1860s and 1870s, national mountaineering clubs emerged with regional branches. Class-based exclusivity was to some extent mitigated by shared nationalist aspirations and shared affinities for local mountain landscapes. They frequently catered for and in some ways sought to unify different expressions of the outdoor movement, with differing class components and affiliations, such as hiking, mountaineering and skiing.[53]

Whilst valuing exclusivity, British Alpine mountaineers helped to construct a new cultural approach to mountain landscapes. They approached mountain terrain and created imaginary territories, aestheticised and intellectualised places, with the Alps at the apex of an idealised taxonomy of landscapes. 'The construction of a category "mountain" is thus social ... the product of collective imaginations'.[54] For most of the nineteenth century, the mountains and fells of the Lake District were understood by British mountaineers in comparison with the Alps, lesser Forms of the Ideal conception of mountains perhaps, but also as part of a typology of mountain landscapes where pre-existing recreational and touristic paradigms could be adopted and adapted. The emergence of rock climbing in the Lake District in the 1880s and 1890s represented a repurposing of elite cultural models by the regional middle classes, hungry to develop the recreational opportunities of

---

52   Buzard, *The Beaten Track*, p. 81.

53   Hoibian, *Les Alpinistes en France, 1870–1950*; Y. Drouet, 'The CAF and the Borders: geopolitical and military stakes in the creation of the French Alpine Club', *The International Journal of the History of Sport* 22 (2005), pp. 59–69; S. Stumpp, 'Alsatian ski clubs between 1896 and 1914: an exploratory evaluation of the role of employees in the German "sportization" of skiing', *The International Journal of the History of Sport* 27 (2010), pp. 658–74.

54   G. Rudaz, 'Stewards of the Mountains: The Poetics and Politics of Local Knowledge in the Valasian Alps', in Cosgrove and della Dora, *High Places*, p. 149.

the upland hinterlands that the expanding railway network and increasing leisure time brought within reach of the industrial cities of the North. Whilst the Alpine holiday remained an elite preserve, mountaineers could not help but write about their experiences in the middle-class quarterlies, newspapers, journals and books. New recreational paradigms became available to the regional and local middle classes, mediated through print, illustration and later photographs. In the industrial cities of the North, mountaineering began to be cultivated amongst prosperous industrialists and the professional and commercial classes. Whilst the Alpine vacation retained its currency as cultural capital, the regional middle classes began to celebrate the recreational potential of hills closer to home. 'Culture' it has been noted, 'proceeds incrementally, building on whatever was available before, sometimes using a well-tried and established formula, other times innovating radically'.[55] In the Lake District in the 1880s and 1890s the conditions emerged that fostered radical innovation. The athletics boom in the public schools and the growth of the gymnastics, body cultivation and physical fitness movements empowered new approaches to the physical challenge of mountaineering. Widening class participation began to undermine the gentlemanly codes that governed mountaineering. Pressure began to build for voluntary associations that would better represent the aspirations of the regional outdoor movement. Rock climbing gradually developed as a distinctive approach to the mountains and crags of Britain, and the Lake District landscape played a central role in shaping this emergent new recreational paradigm.

The growing popularity of mountaineering literature and imagery fostered localised approaches to the regional mountains of the British Isles. In the nineteenth century vigorous walking and scrambling on the Lakeland fells developed as a leisure pursuit amongst the middle classes in Cumberland and Westmorland, with its own cultural traditions and equipment. To walk in the fells and explore your native county was a celebration of regional distinctiveness. To be a good fellsman or fellswoman was a form of territoriality that expressed local exceptionalism and celebrated regional identity. This was framed within wider historical self-conceptions of distinctiveness: individual autonomy, personal liberty, nobility and a love of freedom were said to be in part racially defined, the Scandinavian origins of the Cumbrian yeoman and statesman being much discussed regional characteristics. Gradually, as Alpine mountaineers began to explore the recreational possibilities of Britain's upland landscapes, visiting the Lake District first on family vacations, then as training for the Alps, they increasingly came into contact with local fellsmen; important climbing partnerships emerged. Thus the Penrith-based corn dealer George Seatree, and the Lorton-based land agent and farmer, J.W. Robinson played critical roles as mentors, climbing partners and friends to early

---

55 D. Sassoon, *The Culture of the Europeans from 1800 to the Present* (London: Harper Press, 2006), p. xvi.

rock climbers.[56] This 'fellowship of the hills' and the liminality of mountaineering culture did much to engender a new kind of freedom: a temporary release from the suffocating restrictions and limitations of the class system. The discourse around the nobility of the Cumbrian statesman in many ways echoed Rousseau's idealisation of mountain virtue, satisfying an upper-middle class desire for *völkisch* authenticity. Against a backdrop of increasing concern over racial degeneration, this racially inflected idealisation of the vitality and nobility of mountain stock is clearly seen in the eulogies for J.W. Robinson. Delivering an address on the occasion of a dedication of a memorial to Robinson on Pillar Rock on 13 June 1908, the Yorkshire solicitor Cecil Slingsby noted:

> Our dear old friend possessed in a high degree all the best characteristics of the north country yeoman, the back-bone of our race. In many cases, and most certainly in that of Robinson, these northern yeomen are the descendents of the Norse "bonder". John Robinson was essentially of Scandinavian ancestry, and I have often called him a British Norseman. If I could have paid him a higher compliment I would have done so.[57]

George Seatree, writing Robinson's obituary in the first volume of the *Fell and Rock Climbing Club Journal* delivered his own perception of Robinson, less insistent on racial lineage but still focussed on regional exceptionalism and local identity, mentioning Robinson's lifelong adherence to the Society of Friends and the Pardshaw Meeting, and his family roots as freeholders.[58] Gradually these interlocking senses of identity, articulated within the wider outdoor movement, shaping the cultural landscape of the Lake District. The middle class Liberal concern for individual liberty found its physical embodiment in the idealised Cumberland statesman. This was reinforced by wider regional affiliations. As the old county boundary of Lancashire went up to the Three Shire Stone on Wrynose Pass, the Coniston fells and the southern Lakes were Lancastrian. Far from being 'other' or distant, the fierce county loyalties of mountaineers from the industrial cites of Lancashire saw the southern Lake District as 'our' hills. County loyalty and pride was a major component of regional self-identity. When W.G. Collingwood wrote his influential 1902 guide to the region he called it *The Lake Counties*, not 'The Lake District'. With the growth of the industrial centre of Barrow-in-Furness, this would reinforce the growing sense of entitlement

---

56   Hankinson, *A Century on the Crags*, pp. 28–40; C. Mill, *Norman Collie: A Life in Two Worlds. Mountain Explorer and Scientist, 1854–1942* (Aberdeen: Aberdeen University Press, 1987), pp. 35–40.

57   M. Waller, *A Lakeland Climbing Pioneer: John Wilson Robinson of Whinfell Hall* (Carlisle: Bookcase, 2007), p. 139.

58   G. Seatree, 'In Memoriam: John W. Robinson', *Fell and Rock Climbing Club Journal* 1, 1 (1907), p. 4. http://www.frcc.co.uk/archive/1907-1919/V1-1.pdf (accessed 2 June 2011).

and legitimacy amongst lower-middle class and working class climbers like George Basterfield and A.H. Griffin after the First World War. This developing sense of regional identity, of the uniqueness of the industrial cites of the north and their relationship to their upland hinterlands, did much to break down class barriers. Specific county identities were also understood in the wider context of Northernness. Slingsby emphasises the 'north country yeoman', his cultural reference point being further north still, in Scandinavia. Slingsby, described by British mountaineering historians as 'the father of Norwegian mountaineering', had climbed in Norway since 1872 and did much to popularise it to mountaineers as a 'northern playground' to rival the Alps.[59] Directly influenced by Ambleside resident Harriet Martineau's writings on Norway, Slingsby was undoubtedly also influenced by W.G. Collingwood's scholarly, popular and fictional explorations of the Lake District's Scandinavian heritage.[60] Teutonic and Norse legend, literature and culture were becoming increasingly important to the regional middle classes in the north of England. Offering cultural reference points that were cosmopolitan and international, they bypassed national and imperial discourses on identity and enabled the construction of a regional sense of exceptionalism and a localised sense of Englishness.[61]

Changing recreational attitudes to the mountains, fells and crags of the Lake District were slow to emerge however. A writer in the *Penny Magazine* of 1837 could not persuade his Wasdale-based walking guide to take him up Broad Stand and this attitude remained typical of the Dalesmen who guided tourists over mountain paths.[62] The first professional climbing guide was only employed in 1901 by the new proprietor of the Wastwater Hotel in Wasdale, J. Ritson Whiting, in an attempt to maintain his climbing clientele. His choice of guide, Joseph Gaspard from the Dauphiné, suggests an effort to emulate the Alpine mountaineering tourist experience but also a lack of indigenous candidates who would be willing to trade amateur status as mountaineers for the indignities of oiling and nailing their clients' boots.[63] Gaspard spent the lucrative summer months of June, July and August back

---

59   C. Slingsby, *Norway: The Northern Playground* (Findon, Aberdeenshire: Ripping Yarns, 2003 – first published Edinburgh: David Douglas, 1904).

60   J.S. Dearden, 'Collingwood, William Gershom (1854–1932)', *Oxford Dictionary of National Biography* (Oxford University Press, 2004); online edn, Sept. 2010, http://www.oxforddnb.com/view/article/39918 (accessed 2 June 2011) doi:10.1093/ref:odnb/39918; W. Rollinson, 'William Gershom Collingwood, 1854–1932', in W.G. Collingwood, *The Lake Counties* (London: J.M. Dent & Sons, 1988), pp. viii–x.

61   Westaway, 'The German Community in Manchester'; P. Readman, 'The place of the past in English culture c.1890–1914', *Past and Present* 186, 1 (February 2005), pp. 147–99.

62   R.W. Clark and E.C. Pyatt, *Mountaineering in Britain: A History from the Earliest Times to the Present Day* (London: Phoenix House, 1957), p. 25.

63   Hankinson, *A Century on the Crags,* p. 66.

in his native Dauphiné.[64] Exploration was generally sporadic and casual, with some notable exceptions. Broad Stand was ascended by the Swiss born Lake District resident C.A.O. Baumgartner in 1850. He also climbed in Snowdonia and was 'one of the first Alpine climbers who deliberately sought out and climbed the individual rock features of the British hills – not the hills themselves, but the individual ridges, buttresses, crags, and other rock features'.[65] Broad Stand was ascended again in 1857 by Prof. John Tyndall, the pioneer of rock climbing in the Alps, who described it as 'a pleasant bit of mountain practice and nothing more'.[66] Drawn to vacation in the Lake District by its Romantic and literary associations, Alpine mountaineers began to concede that vigorous walking and scrambling up rocks and gullies might offer some 'pleasant relaxation which had a direct bearing on their activities in the greater mountains of the world'.[67] For a while in mid-century Pillar Rock in Ennerdale became the focus of attention as its summit was only attainable by moderate scrambling and rock climbing, Leslie Stephen ascending it in 1865 and again in 1872.[68]

Gradually more organised approaches to the mountains of England and Wales began to emerge from within the Alpine Club. C.E. Matthews of Birmingham, an original member of the Alpine Club, visited Pen-y-Gwryd in Snowdonia in January 1870: 'At that time of year', wrote Matthews, 'we were quite certain of having the inn to ourselves; we almost always found snow on the hills',[69] an important consideration for those wanting to practice their mountaineering craft in preparation for the Alps. Matthews went on to form his small group of friends into the Society of Welsh Rabbits, dedicated to spending annual meets around Christmas time at the Pen-y-Gwryd and climbing throughout the British Isles. Perhaps the individual who did the most to reposition the Alpine Club's attitude to the mountains of Britain was the Manchester merchant Eustace Hulton. Between 1875 and 1883 he organised the annual Alpine Club Dinner and Meet each spring, in either the Lake District or North Wales. These events usually involved strenuous walking and scrambling, with an eye for mountaineering challenges, particularly if there was snow on the ground. Cecil Slingsby, delivering a paper to the Alpine Club in April 1886 exhorted its members to 'not neglect the Lake District, Wales and Scotland, whilst we are conquerors abroad'.[70] Even so, as late as 1894, according to Prof. Norman Collie, the bulk of the Alpine Club membership remained in ignorance of the mountaineering and climbing potential of the Lake District and dismissive of

---

64   N. Price, *A Vagabond's Way: Haphazard Wanderings on the Fells* (London: George Allen & Unwin, 1914 – 3rd edn 1937), pp. 72–3.

65   Clark and Pyatt, *Mountaineering in Britain*, p. 22.

66   Ibid., p. 25.

67   Ibid., p. 26.

68   Ibid., p. 24; Hankinson, *A Century on the Crags*, pp.15–21.

69   Clark and Pyatt, *Mountaineering in Britain*, p. 29.

70   Ibid., pp. 42–3.

the emerging sport of rock climbing.[71] Hulton's biggest influence was to encourage the mountaineering aspirations of a number of Lancashire industrialist families in the 1870s and 1880s, notably the Pendleburys of Liverpool and the Pilkingtons and Hopkinsons of Alderley Edge. Keen Alpinists, they were pioneers of climbing in the Lake District and were seminal in supporting the nascent outdoor movement in north-west England. Charles Pilkington's presidency of the Alpine Club 1896–1898, and his tireless lecturing on mountaineering to rambling groups and climbing clubs, did much to narrow the social distance between elite clubs like the Alpine Club and the new regional clubs.

By the 1880s we can identify the emergence of a highly innovative mountaineering and climbing culture in and around Wasdale Head in the Lake District. Centred on the Wastwater Hotel and the guest house at Row Head, the isolation, difficulty of access and lack of accommodation in the valley had the effect of concentrating tourist provision for visiting mountaineers, climbers and walkers. As late as the early twentieth century the western Lakes were indeed 'unique in their continuing isolation from the main tourist centres', 'beyond the reach of the day excursion from Keswick, except for the most determined'.[72] Most visitors came from the West Cumberland seaside resorts, arriving by train and travelling along the carriage roads to Wasdale Head. For visitors intent on climbing, the main mode of access to Wasdale was walking. Vigorous pedestrians like Leslie Stephen, enjoying a short break in the Lakes, could take the train from Euston on a Saturday in April 1875 and arrived at Furness Abbey at teatime. On Sunday he proceeded to Foxfield Junction, embarking on the Coniston Railway to Coniston, arriving at about midday and setting off to walk to Wasdale Head: 'Thence we had to walk here crossing three ridges on the way. We managed the first two pretty well but J.W. became uncommonly tired over the last & I had enough of it – my knapsack being very heavy & nearly cutting my shoulder in two'.[73] The following Wednesday he walked back to Coniston to catch the train home. Wasdale was 'assuredly the core of the mountains' according to the actress Nancy Price. Arriving there on foot in 1914 she noted:

> I think one should always enter Wasdale as daylight fades. I know that I felt everything to be essentially right as we scrunched down the Sty Head Pass in the gathering dusk towards the Inn – I beg its pardon, 'The Wasdale Head Hotel'.[74]

This hotel had been established at Wasdale Head by William Ritson, who had inherited the farm at Row Foot from his grandfather, Bill Ritson. In 1856 'in

---

71 Mill, *Norman Collie*, p. 36; Hankinson, *A Century on the Crags*, p. 35.

72 J.D. Marshall and J.K. Walton, *The Lake Counties from 1830 to the Mid-twentieth Century* (Manchester: Manchester University Press, 1981), pp. 190–91.

73 J.W. Bicknell (ed.), *Selected Letters of Leslie Stephen: Volume 1, 1864–1882* (London: Macmillan, 1996), pp. 153–55.

74 Price, *A Vagabond's Way*, p. 65.

response to the increasing number of visitors to the valley, he applied for a licence to sell alcohol' and at the southern end of the farmhouse 'added a small wing to provide accommodation' which became known as the Huntsman's Inn. Ritson would 'occasionally act as a mountain guide for visitors, but had no interest in rock climbing and usually tried to dissuade tourists from attempting it'.[75] In the 1870s Tom and Anne Tyson established a teetotal guesthouse in the adjacent farm at Row Head which established a 'reputation for good home cooking'.[76] In 1868 William Ritson's farmhouse at Wasdale Head was said to offer '"clean and comfortable" lodgings at Wasdale Head for ten or twelve people at a time'[77] but it had a reputation for being 'raucous and unpredictable'. Ritson himself was something of a Cumberland celebrity: a 'keen foxhunter, champion wrestler, excellent storyteller and shrewd businessman'.[78] In 1875 Leslie Stephen described Ritson's as 'the queerest of little places ... The Ritsons [the landlords] sit in the kitchen all day & callers walk in & out. The roof is timber & hung with flitches of bacon & guns [?]; & I knock my head against the beams in the passages if ever I don't look out'.[79] In 1879 'Will Ritson and his wife Dinah retired from the Huntsman's Inn and Daniel Tyson, involved in running the Row Head guesthouse, took over the lease',[80] changing the name to the Wastwater Hotel. Slightly higher up the valley, the farm at Burnthwaite also took in guests. In 1935 Dorothy Pilley recalled her first visit to Wasdale *c*. 1910–1920 and the changes that had come over Burnthwaite:

> Baths and modernity have come to it. Then, one sat in a dark, cosy little room and ate in a narrow whitewashed cell, which I believe had once been a dairy. If you were much favoured, old Mrs. Wilson would let you sit and gossip in the kitchen.[81]

A visitors' book was kept at the Huntsman's Inn from 1863 and early records of mountaineering exploits were transcribed from this to the Wasdale Climbing Book begun in 1890 at the Wastwater Hotel.[82] Along with the visitors' book kept by the Tysons at Row Head, from which extensive accounts have been

---

75   M. Cocker, *Wasdale Climbing Book: A History of Early Rock Climbing in the Lake District based on contemporary accounts from the Wastwater Hotel, 1863–1919* (Glasgow: Ernest Press, 2006), p. 21.

76   Ibid., p. 22.

77   Marshall and Walton, *The Lake Counties*, p. 190.

78   Cocker, *Wasdale Climbing Book*, p. 21.

79   Bicknell, *Selected Letters of Leslie Stephen*, p. 154.

80   Cocker *Wasdale Climbing Book*, p. 22.

81   D. Pilley, *Climbing Days* (London: The Hogarth Press, 1989), p. 58.

82   Now part of the Fell and Rock Climbing Club archive, Lancaster University Library. See Cocker, *Wasdale Climbing Book*, p. 5.

published for the years 1876–1886,[83] they form an invaluable historical record of early mountaineering and rock climbing in the valley. Catering for tourists, walkers, mountaineers and climbers, these visitors' books provide evidence of visits by the professional classes from northern industrial towns but also indicate a high proportion of scholars, academics and dons from the public schools and universities, corroborating Noel Annan's statement about the attraction of mountains and mountaineering to intellectuals. Wasdale Head seems to have been a popular destination for academic reading weeks, offering cheap accommodation, inspirational landscapes, vigorous exercise and few distractions. Sufficiently off the beaten track to deter the trippers that were clogging up Bowness and Keswick and mountainous enough to satisfy aesthetic and athletic requirements, Wasdale combined rustic authenticity and a degree of remoteness and exclusivity. Lehmann Oppenheimer, writing in the *Climbers' Club Journal* of 1899 in an article entitled 'Wastdale Head at Easter' could write:

> there are few spots of equal beauty so undisturbed by traffic, noise, and other accompaniments of a civilization to which too many holiday resorts have fallen victims. As yet no signboards point the way to the best scenery, the waterfalls are unenclosed, and the mountain sides undesecrated by railways.[84]

John Stogden, a Harrow Schoolmaster and one of the pioneers of guideless climbing in the Alps, was a regular guest at the Wasdale Head Hotel, the first record of him visiting being in January 1870. He was 'one of the first to draw the attention of his fellow-Alpinists to the opportunities offered by hills nearer home'.[85] His account in the Alpine Club's *Alpine Journal* of the winter ascent of South Gully on Bowfell was 'the first time the journal had published anything relating to climbing in Britain', this a full thirteen years after the founding of the Alpine Club.[86] The Row Head guesthouse was also a popular destination for university reading parties. Charles Cannan, the Dean of Trinity College, Oxford, introduced the likes of C.E. Montague, A.E.W. Mason and Arthur Quiller Couch to the delights of Mrs Tyson's 'sweet mountain of mutton and Mr. Pendlebury's pudding ... a delicious compound of farm milk, tapioca and raisins'.[87] Reading parties taken at Easter or Whitsun or between academic terms remained short but could be prolonged in the summer vacation. In 1881 the twenty-two year old Walter Parry Haskett Smith spent two months in Wasdale as part of a university

---

83   H. and M. Jackson, *Lakeland's Pioneer Rock-Climbers: Based on the Visitors Book of the Tysons of Wasdale Head, 1876–1886* (Clapham, North Yorkshire: Dalesman Books, 1980).

84   L.J. Oppenheimer, *Heart of Lakeland* (Glasgow: Ernest Press, 1988), p. 15.

85   Hankinson., *A Century on the Crags*, pp. 21–2; M. Cocker, *Wasdale Climbing Book*, p. 2.

86   Cocker, *Wasdale Climbing Book*, p. 18.

87   Hankinson, *A Century on the Crags*, p. 48.

reading party, staying at Row Head for the first month. He was back in 1882 with his brother Edmund for nine weeks.[88] For the mountaineering tourist, particularly those with limited vacation time and on a budget, Cumberland offered a cost effective and time effective alternative to Switzerland. Writing *c.*1898–1899, the London school teacher O.G. Jones noted:

> We cannot conveniently reach Switzerland at every season of the year. At Christmas and Easter it is entirely barred to most people. The expense of foreign travel is a consideration, and the question of length of holiday is rarely negligible. Cumberland can be reached in a night from London; the district is an inexpensive one for tourists. ... Personally I should always go to the high Alps when the chance offered itself, but Cumberland serves remarkably well to allay the desire for mountain air and vigorous exercise when Switzerland is out of the question.[89]

The intensity of mountaineering exploration in the Lake District began to pick up in the 1880s and news of this began to be reported in popular periodicals. After spending mid-August 1884 at Row Head, climbing with large parties that included J.W. Robinson, the London journalist C.N. Williamson was able to publish what is arguably the first climbing guide to the region, as an article entitled 'The Climbs of the English Lake District' in the Magazine *All The Year Round.*[90] It is instructive that Williamson's piece did not find its way into the quarterlies like the *Cornhill* or the *Nineteenth Century*, where so many Alpine Club members chose to publish Alpine accounts alongside the higher journalism, but in a middlebrow publication, edited by Charles Dickens Jnr, aimed at popular audiences. Williamson's article presaged the explosion of mountaineering articles and stories that appeared in the 1890s in popular monthlies like George Newnes' *Strand* and *Wide World Magazine*, appealing to a wider class of readership.

The event that did the most to publicise climbing in Wasdale however, occurred in June 1886, when W.P. Haskett Smith established an epochal climb that most mountaineering historians use as their datum for the birth of rock climbing in the British Isles, climbing the Wasdale Crack on Napes Needle. The climb, described by Alan Hankinson as 'arguably the most significant short climb ever made',[91] represented a new level of technical difficulty on steep open rock,

---

88   Cocker, *Wasdale Climbing Book*, p. 23.

89   O.G. Jones, *Rock Climbing in the English Lake District*, 2nd edn (Didsbury: E. J. Morten, 1973), pp. lvii–lviii.

90   Cocker, *Wasdale Climbing Book*, p. 24; Jackson and Jackson, *Lakeland's Pioneer Rock-Climbers*, pp. 138–9; S. Reid, 'A History of Lake District Climbing Guidebooks', *FRCC Centenary Journal* (2006), p. 873: http://manage.hotscot.net/customer_images/ FA00F036-E183-44C7-A38C-40C3F0D4FE18/archive%20journals/frcc%20guides%20 history%20frcc%20j%202006.pdf (accessed 2 June 2011).

91   Hankinson, *A Century on the Crags*, p. 40.

requiring gymnastic ability and athletic conditioning, and was a celebration of the kinaesthetic joy of climbing and a youthful statement of indifference towards much that the mountaineering establishment held dear. It is clear evidence of the culture of mountaineering beginning to hybridise, with new sub-cultures emerging. News of the ascent and the iconic image of the Needle shaped perceptions of the emerging sport of rock climbing, being reported in the *Pall Mall Budget* of 1890. In the early 1890s the sight of Professor H. Dixon's photograph of climbers on Napes Needle 'in a shop window in the Strand' inspired O.G. Jones to start climbing in the Lake District.[92] The Abraham Brothers of Keswick, photographic entrepreneurs and mountaineers, sold photos of Napes Needle to tourists (Figure 8.1). Being the first to introduce the scenic picture postcard to Britain,[93] having seen them on sale in Zermatt in 1898, the trade in cheap images of the Lake District landscape and of climbing and mountaineering became a core part of their business. On a subsequent trip to the Haute Savoie, George Abraham 'was amused to discover one of their views of climbers on Napes Needle on sale in the shops of Chamonix, masquerading as an Alpine pinnacle called the "Aiguille de la Nuque"'.[94] By 1902 W.G. Collingwood could state that Napes Needle 'is pictured on all the posters of Lake District attractions',[95] part of the distinctive landscape repertoire of images that the tourist industry sought to promote. The Abrahams filmed an ascent of Napes Needle, possibly as early as 1913, and in the early 1920s produced a 'short travelogue-type film – a Pathé Review with printed captions' showing George Abraham leading the Arête route on Napes Needle.[96]

What emerged in the decades up to the First World War at Wasdale was a highly distinctive climbing and mountaineering milieu that was recognised as such by contemporaries.[97] Wasdale Head cultivated the new sport and became a

---

92   A. Hankinson, *Camera on the Crags: a portfolio of early rock climbing photographs by the Abraham Brothers* (London: Heinemann, 1975), p. 3.

93   I believe this refers to Orell Fussli's Photochrome postcards, marketed by Photoglob in Zurich. Having perfected his new colour process in 1889, Fussli won first prize at the Exposition Universelle in Paris in 1900. See M. Walter and S. Arque, *The World in 1900: A Colour Portrait* (London: Thames & Hudson, 2007), p. 13.

94   Hankinson, *Camera on the Crags*, p. 20.

95   Collingwood, *The Lake Counties*, p. 62.

96   Hankinson, *Camera on the Crags*, p. 27. See also the BBC Cumbria webpage, 'Lakeland Life on Camera, 2008', featuring the documentary *Inside Out: The Abraham Brothers*:   http://www.bbc.co.uk/cumbria/content/articles/2008/10/13/abraham_bros_io_feature.shtml (accessed 7 June 2001).

97   Oppenheimer, *Heart of Lakeland*; G.S. Sansom, *Climbing at Wasdale Before the War* (Castle Cary, Somerset: Castle Cary Press, 1982); A.H. Griffin, *The Coniston Tigers: Seventy Years of Mountain Adventure* (Wilmslow: Sigma Leisure, 2000), pp. 43–57; Price, *A Vagabond's Way*, pp. 65–95.

Figure 8.1   The prominent pinnacle of Napes Needle on the side of Great
              Gable, photographed by George and Ashley Abraham
Source: © Susan Steinberg

forcing ground for harder rock routes as well as winter mountaineering routes.[98] The shortage of accommodation in the valley led to intense overcrowding, particularly at Easter and Whitsun, W.G. Collingwood noting in 1902 that 'to enjoy this neighbourhood you must go there out of season; that is to say, any time except Easter and summer vacations'.[99] Class distinctions based on hotel status or room tariffs became impossible to maintain. There was simply no choice and more than a touch of the 'take-it-or leave-it' approach that Marshall noted marked out the Cumbrian innkeeper, 'a man of considerable independence of attitude and outlook ... prone to think that others should enjoy without complaint the food and linen to which he himself was accustomed'.[100] As the mountaineers virtually took over the hotel at certain times of year a highly playful and anarchic culture descended. Oppenheimer's essay 'Wastdale Head at Easter' describes catching the night train from Manchester and walking the twelve miles from the coast to Wasdale Head. Enquiring of Mr. Tyson about a room he is told that 'every room in the house was taken two or three weeks ago; there's some folks sleepin' i'the smoke-room and some i'the barn – not a bed to spare nowhere, and they are full up at Burnthwaite too'.[101] Far from being trippers from Barrow and Whitehaven, Tyson informed him that few were 'fro these parts: mostly they're from all over England – London and Oxford, Yorkshire and Manchester'.[102] Professors, poets, bohemians, undergraduates, solicitors, the crowd Oppenheimer describes is cultivated, well travelled and not unfamiliar with the Alps. There is undoubtedly a preponderance of 'higher professionals' in Oppenheimer's account.[103] But we know that regional industrialists and businessmen were not underrepresented and that many of the academics from the regional red brick universities were chemists and engineers with strong links to industry. Further down the social scale this milieu was open to tradesmen like the Abrahams of Keswick (photographers), Oppenheimer (mosaic flooring) and the lower-middle class teachers like O.G. Jones. Contemporary mountaineers remarked on the democratic nature of these gatherings, where class distinctions were discounted. Describing her first night at the Wastwater Hotel, Nancy Price, having overcome the embarrassment of being the only woman there and having the whole room stand when she came in to dinner, described the conviviality:

---

98   Hankinson, *A Century on the Crags*, pp. 41–87; M. Cocker and C. Wells, 'Winter Climbing in the Lake District 1870–1941: a list of first recorded ascents and early attempts', *FRCC Journal* (2002): http://manage.hotscot.net/customer_images/FA00F036-E183-44C7-A38C-40C3F0D4FE18/archive%20articles/winter%20climbing%20history.pdf (accessed 2 June 2011).

99   Collingwood, *The Lake Counties*, p. 64.

100   J.D. Marshall, *Old Lakeland: Some Cumbrian Social History* (Newton Abbot: David & Charles, 1971), pp. 168–9.

101   Oppenheimer, *Heart of Lakeland*, p. 16.

102   Ibid., p. 17.

103   R. McKibbin, *Classes and Cultures in England, 1981–1951* (Oxford: Oxford Univerity Press, 1988), p. 46.

Everybody talked to everybody else – and what talk! I fancied myself at Zermatt. Ropes and axes, couloirs, arêtes, chimneys. … That first night at Wasdale there sat down to dinner, two clerks from Barrow, an Oxford undergraduate … two keen young Americans … a well known author, and a gentleman who regaled us with tales of the good old days.[104]

In this almost entirely male atmosphere, the ludic nature of recreation came to the fore. Training games like the Billiard Room Traverse, and the Barn Door Traverse and upside-down bouldering on the Y boulder in Mosedale demonstrated the intense physicality of this new culture, its inventiveness and playfulness, sub-cultural innovations that prefigure today's climbing walls and bouldering scene. The climbing culture at Wasdale witnessed the sportisation of mountaineering, the gradual working out of rules and codes of behaviour very different from what had gone before, less governed by external literary, aesthetic and scientific criteria, more concerned with things of interest to a small coterie within the sport: issues of what constituted acceptable and unacceptable climbing ethics, techniques and equipment. And this was reinforced through the development of club rituals, songs and the climbing club journals that were starting to emerge. If 'one of the most important characteristics of play was its spatial separation from ordinary life',[105] then the remoteness and spatial separation of Wasdale did much to foster this playful innovation. Many climbers were conscious that travel began the process of separation, often meditating on the transition from the smogs of industrial Britain to the clean air and renewed clarity of the fells. In the night journeys by train, or huddled around the fire in the early hours of the morning at Penrith station, they began to put on new selves. Geoffrey Winthrop Young noted that the journey served to:

cover a similar change in myself, from an evening self laboriously constructed under a thousand pressures to do a hundred civilized tricks, to a spontaneous self that came of itself, with the winds and the height and the rapture of morning and movement. I was not introspective as a boy, but I must quickly have recognized that some sort of transition would take place; for I used to wait to change into mountain clothes until I was on the neutral territory of the train, so that the pleasanter self which waited for me somewhere about the thousand-foot level might not return to life in inappropriate trousers.[106]

The democratic culture on display at Wasdale and the 'return to life' that mountaineering tourism offered led many contemporary commentators and social

---

104   Price, *A Vagabond's Way*, pp . 69–70.

105   J. Huizinga, *Homo Ludens: A Study of the Play Element in Culture* (London: Temple Smith, 1970), p. 38.

106   G.W. Young, *On High Hills: Memories of the Alps*, 5th edn (London: Methuen, 1947), p. 24.

theorists to place great hope in the outdoor movement. C.E. Montague, chief leader writer at the *Manchester Guardian,* heavily promoted the outdoor movement and mountaineering in particular between 1890 and 1926, incorporating it into a critique of society based on New Liberal social theory. He saw it as a panacea for the ills of industrial society, a new space where class based politics would be replaced by rational, classless recreation, which celebrated regional identity and reunited socialist and liberal elements in the progressive movement. He appealed to the regional middle classes to take a more active role in the social leadership of the outdoor movement. Gradually, clubs that focussed on mountaineering and climbing in the British Isles emerged. In England the first to be established were clustered around the Pennines and in industrial Lancashire: The Yorkshire Ramblers Club (f. 1892), The Kyndwr Club (f. 1895), The Manchester based Rucksack Club (f. 1902), the Derbyshire Pennine Club (f. 1906), and the Liverpool-based Wayfarers Club (f. 1906).[107] The Climbers' Club was founded in London in December 1897 to bring together those interested in 'mountaineering in England, Ireland and Wales', evidence, its founders suggested of something 'hitherto only half suspected – of a large body of British climbers ready for an association from which the organisation and development of their sport might be looked for'.[108] At its inception the Climbers' Club was careful to position itself, stating 'the Club will be in no sense antagonistic to any existing institution', respecting the territorial sensibilities of the Scottish clubs but also ongoing accusations of vulgarisation from the Alpine Club.[109] The Lake District itself was among the last in this movement to establish regional mountaineering clubs, the Fell and Rock Climbing Club being formed 1907. Whilst its committee members were all resident in and around the Lake District it sought to boost its social and sporting legitimacy by inviting noted Alpinists from the region who also had strong Lake District climbing credentials, like Slingsby, Collie, Geoffrey Hastings and Charles Pilkington, to be honorary members.

Perhaps the most striking thing of all about the climbing culture that emerged in the period in the Lake District was its informality, emerging largely without formal voluntary associational structures. Even after the foundation of the regional clubs, bourgeois-romantic individualism remained in tension with further institutionalisation, deeply ambivalent towards the mass appropriation of the hills that would be such a feature of the post-war period. The production of guidebooks was left to private individuals, and the First Series of F.R.C.C. guidebooks only

---

107   F.D. Smith, *A Brief History/Introduction to the Yorkshire Ramblers' Club* (2011): http://www.yrc.org.uk/content.aspx?Group=about&Page=history%20overview (accessed 2 June 2011); J.P. Craddock, *Jim Puttrell: Pioneer Climber and Cave Explorer* (Kibworth Beauchamp: Matador, 2009), pp. 33–4, 178; K. Fyles and B. Stroude (eds), *100 Years of Wayfaring, 1906–2006* (Liverpool: The Wayfarer's Club, 2006).

108   Bryan, 'The Formation of the Climbers' Club', p. 4.

109   Ibid.; L.J. Oppenheimer, 'Correspondence, Manchester, March 16th 1901', *Climbers' Club Journal* 3, 11 (1901), pp. 133–5.

began to appear under R.S.T. Chorley's editorship, G.S. Bower's *Doe Crag* appearing in 1922.[110] Whilst the Rucksack Club had opened its first hut at Cwm Eigau in North Wales in 1912, followed by the Climbers' Club hut at Helyg in Ogwen in 1925 and the Rucksack Club's Tal-y-Braich Uchaf, Ogwen, 1927, the first climbing club hut in the Lake District did not open until the Wayfarers' Club opened the Robertson Lamb Hut in Langdale in 1930. The Coniston Tigers assembled their own hut (an old garage) behind Coniston Old Hall in 1931.[111] The F.R.C.C. opened its Brackenclose hut in Wasdale in 1937.[112] Before the inception of dedicated climbing club huts the location, provision and nature of tourist accommodation proved hugely influential in shaping the emerging sport, its rituals and traditions.

Mountaineers and climbers in the pre-war period often had multiple club affiliations, expressing different localised identities, but the new regional clubs remained firmly middle-class voluntary associations, often negotiating private access agreements with landowners in the Pennines in a mutually beneficial compact that kept access to Pennine grouse moors tightly restricted. Whilst the mountains of Snowdonia and North Wales witnessed similar developments in the sportisation of mountaineering, their cultural importance to recrudescent Welsh nationalist identity complicated attempts to appropriate them as cultural landscapes and tourist destinations. By 1914 the preponderance of climbers and mountaineers from England looked towards the Lake District as the culturally most significant British mountain landscape, an evaluation reinforced by its Romantic and literary association as well as its relative accessibility. Its cultural significance as the 'birthplace of rock climbing' was already firmly established. Its increasing accessibility meant that climbers from the industrial cities of the North could apply climbing techniques honed on the gritstone edges of the Pennines to mountain crags, leading to acceleration in the severity of climbs undertaken. The pre-war high watermark in terms of difficulty was reached in April 1914 with Siegfried Herford's, George Sansom's and C.F. Holland's ascent of Central Buttress on Scafell. In choosing to find a suitable way to memorialise Herford and other mountaineers who had died in the Great War in 1919, the F.R.C.C. elected to purchase Great Gable, eventually unveiling a bronze memorial plaque on its summit in 1924.[113] No better cenotaph than the mountain itself was required to express the complete self-identification shared by English mountaineers with the Lake District landscape. Sacralising the landscape added to its cultural valency. The commonality of sacrifice in the Great War had reinvigorated arguments for

110   Reid, 'A History of Lake District Climbing Guidebooks', p. 854.

111   Griffin, *The Coniston Tigers*, p. 63.

112   I am indebted to Douglas Hope for this information on club huts, forming part of his PhD research at the University of Cumbria: D.H. Hope, 'The Changing Role and Influence of the Organisations that Pioneered the Provision of Recreational and Educational Holidays for Working People in the English Lake District, *c*.1930–2004' (forthcoming).

113   Hankinson, *A Century on the Crags*, p. 103.

the restorative nature of upland landscapes and renewed arguments for access to mountains. In donating the land to the National Trust the F.R.C.C. looked to the future of widened and more democratic access that was to be such a feature of the outdoor movement in the 1920s and 1930s. At the dedication ceremony on 8 June 1924, Geoffrey Winthrop Young provided a poetic distillation of all the many meanings that mountains signified to mountaineers and climbers: 'Upon this mountain we are met today to dedicate this space of hills to freedom'.[114]

114   A. Hankinson, *Geoffrey Winthrop Young: Poet, Mountaineer, Educator* (London: Hodder & Stoughton, 1995), p. 236.

Chapter 9

Sport, Tourism and Place Identity in the Lake District, 1800–1950

Mike Huggins and Keith Gregson

Modern studies of sport tourism are present-centred, their theory and methodology dominated by sports marketing and sports management disciplines, and focusing on a multitude of activities, destinations and site that reflect the economic importance of sport in contemporary society. In Britain, by the 1990s, sport accounted for approximately 2 per cent of all workers and contributed substantially to Britain's G.D.P.[1] About 20 per cent of trips in Britain then related to sports participation and about 50 per cent had a sports component.[2] A globally popular mega-event like the 2012 Olympics is expected to generate substantial domestic and international tourism turnover.[3] Governments, tourist boards and private industry all now commercially exploit sports tourism, and there is much statistical data available to support researchers' neat theoretical constructs. Definitions of 'sports tourists' vary but researchers largely use three categories: high performance individuals and teams, 'committed' sporting practitioners, and travelling spectators or fans, all visiting 'modern' sporting events.[4] Some studies further distinguish between day-trippers ('sports excursionists') and 'sports tourists' staying overnight.[5] There are other micro-definitional discriminations, between 'sport as competition' and 'sport-as-play',[6] and between trips with sport as the primary purpose and holidays with sport as a secondary or incidental purpose.[7]

---

1 M. Johnes, 'Sports Industry', in R. Cox, G. Jarvie and W. Vamplew (eds), *Encyclopaedia of British Sport* (Oxford: ABC-Clio, 2000), p. 373.

2 Australian Department of Industry, Science and Resources, *Towards a National Sports Tourism Strategy* (Canberra: DISR, 2000).

3 J. Horne and W. Manzenreiter, *Sports Mega-Events: Social Scientific Analyses of a Global Phenomenon* (London: Wiley-Blackwell, 2006).

4 See, for example, A. Tomlinson, *Oxford Dictionary of Sports Studies* (Oxford: Oxford University Press, 2010), pp. 38–9; L.D. Neirotti, 'An Introduction to Sport and Adventure Tourism', in S. Hudson (ed.), *Sport and Adventure Tourism* (Abingdon: Routledge, 2002), p. 2.

5 D. Weaver and M. Oppermann, *Tourism Management* (Brisbane: John Wiley, 2000), p. 28.

6 R. Gruneau, 'Freedom and constraint: the paradoxes of play, games and sport', *Journal of Sport History* 7, 3 (1980), pp. 68–76.

7 J. Standeven and P. De Knop, *Sport Tourism* (Champaign: Human Kinetics, 1999).

But whilst such definitions are possible to apply to current sports tourism, they are less relevant to the analysis of past trends, as this historical study of sport tourism in the Lake District will show. The Lake District, with its continued reliance on natural attractions and its powerful 'place' images, initially created by eighteenth-century art and literature, attracted 'first-class' visitors, many of whom looked to experience something of the aesthetic, Romantic and spiritual uplift associated with its landscape, and the satisfactions of the sublime and picturesque. Before the Second World War, though the range of visitors gradually became more diverse, the attractions of 'modern' sport were rarely a primary purpose of their visits, not least because there was so little of it. Even so the area saw some touring club sides in the summer as early as the late nineteenth century, as cricket tours by Somerset's Castle Cary side in 1897 and Fettes Wanderers, Edinburgh in 1900 illustrated.[8] The Somerset side played eight games in the Lakes, staying at the Old England, Bowness, and at Keswick.

A combination of rugged scenery and sports-related activity was sporting visitors' basic fare, either participating or watching, though bad weather was always problematic.[9] In activities such as fishing, rowing, golf and cycling competition was limited or played no part. Some activities might entail competition, such as the various lake sports of yachting and motor boating or the establishment of water-speed records. Then there were activities depending on wider notions of sport: 'blood sports' like hunting, or hound trailing and sheep dog trials. There were also 'adventure' activities, such as rambling or rock climbing, merging sport and tourism.[10]

As part of their wider holiday experience visitors might also have the opportunity to watch formally organised local sports events. Lakeland annual regattas and some sports meetings were in part spectacles, performances for prize money, laid on as part of the 'festivity', 'entertainment' and 'amusements' for wealthy visitors. Part of Lakeland's appeal was the largely mythical representations of its regional sports, conveying a sense that these reflected a unique community, with a traditional way of life. The novel experience of watching Cumberland and Westmorland sports such as the local back-hold wrestling, fell racing and hound trailing, appeared, at least to naïve, respectable off-comers, to involve apparently 'untainted', 'genuine' competition between 'steadfastly honest', 'noble' dalesmen, even if such images had little objective validity. In tourists' minds they were a welcome contrast to 'professional' commercial sports, though their attendance enhanced the income stream for events, and so indirectly prize money for local competitors. Metropolitan media images the tourists drew on overlooked the reality: that most men competed for prize money more than honour, that leading

---

8   *Westmorland Gazette* (4 August 1900).

9   See J.K. Walton, 'The Windermere tourist trade in the age of the railway', in O.M. Westall (ed.), *Windermere in the Nineteenth Century* (Lancaster: Centre for North-West Regional Studies, 1991), p. 24.

10   Hudson, *Sport and Adventure Tourism*.

wrestling competitors were all semi-professional, that visiting professionals competed in the athletics events and there was substantial betting on results.

Modern theories of sports tourism rightly emphasise the varied motives for sports tourism, and the complex personal economic, intellectual, political or social orientations involved.[11] In the Lakes it is particularly difficult to disentangle tourist motives from the attractions of Lakeland as 'imagined place' with associated 'traditional' sporting events. One *Westmorland Gazette* reporter even claimed that Grasmere Sports, the best-attended sporting event in the region, was 'only an excuse in 99 cases out of 100 for a summer picnic in one of the loveliest spots in England'.[12]

Visiting tourists were alerted to the sporting possibilities of the Lakes not so much through the validation of art and a rich literary tradition as through hotel advertisements, the press, and growing numbers of guidebooks, all useful sources for the historian of sports tourism. Ward Lock, for example started producing travel guides in 1880 as 'pleasant travelling companions' to tourist destinations. They first covered the English Lake District from 1884 and from then to 1937 their 'pictorial and descriptive' guides always contained a standard 'outline guide' of 'sports and amusements', sports-related activities for visitors to engage in, initially listing angling, boating, bowls, golf links, hunting, walking and mountaineering, tennis and winter sports, later adding curling and cycling. Later guidebooks provided details of local sports meetings, foregrounding Grasmere, the fell races, hound trails and Cumberland and Westmorland wrestling, all supposedly 'peculiar to this part of England'.

Recognising this complicated sporting provision and the mixed nature of tourist motivations, the following study takes a broad definition of sporting activity, focusing particularly on sports that traditionally have been perceived by visitors as more distinctively reflecting Lake District identity, and creatively exploiting those few key data sources which can shed limited light on past sports tourism. It is divided into two sections: firstly covering the period up to the First World War and secondly the period from there to the founding of the National Park, a useful closing point since from then on, as the Park's own Annual Reports indicate, more modern forms of sports tourism were emerging. It focuses on the tourists and their experiences, rather than the impact of sports tourism on the Lakes economy.

Sports Tourism before 1914

By the later eighteenth century the Lakes drew small numbers of gentlemanly, leisured, landed or urban professional and mercantile visitors in search of health,

---

11   T. Robinson and S. Gammon, 'Revisiting and applying the sport tourism framework: a question of primary and secondary motives', *Journal of Sport Tourism* 9, 2, (2004), pp. 221–3.

12   *Westmorland Gazette* (20 August 1881).

recreation and exercise. These more adventurous tourists, mostly staying in hotels, filled their days with outdoor activity: excursions, walking, rowing, fishing, painting or watching local summer sports events such as regattas or wrestling.

Railway penetration of the Lakes from the 1840s brought in more middle- and upper class holidaymakers. Numbers reached a peak between July and September, swollen by with second-home owners and limited numbers of generally well-behaved excursionists, especially at Whitsuntide, but growing in numbers from the 1870s. Most visitors stayed one or two weeks or more, in hotels, farmhouses, boarding houses and rented accommodation. Major summer events such as Grasmere Sports and Pooley Bridge, Ullswater Sports increasingly catered for tourist spectators, and regular national media coverage of these sporting 'mega-events' swelled numbers. Yacht racing became popular with wealthy second-home owners; and some visitors took up rambling, climbing or golf as part of their holiday experience. A few came for 'country' sports, as the proliferation of small hunting, shooting or fishing lodges and local newspaper reports indicates, but despite descriptions of grouse shooting on the moors, with names of shooters and their 'bags', or of local hunt meetings, there is insufficient data to distinguish local gentry from guests or seasonal renters.

We have limited insight into numbers of tourists arriving during this period, despite tollgate figures, railway companies' returns and newspaper visitors' lists. A sample study of the latter has suggested that nearly half of all visitors came from the northern counties, especially Lancashire. The rest came from further south, including over 10 per cent from English-speaking countries abroad.[13]

How far did such visitors take part in or watch sporting activities? This is also difficult to assess, though unrestricted access to the fells, large areas of common land and the attraction of Lakeland's 'wilderness' areas made the Lakes a centre for walking, an activity fostered by writers from William Wordsworth onwards. Early tourists enjoyed genteel strolls between viewing points, but excursive walking, drawing on the Romantic poets' example, and engaging with the landscape, was taken up more widely in the second half of the nineteenth century. In 1854 there was still debate in the *Times*'s columns as to whether local guides were necessary when walking across the fells, but increasingly guidebooks and maps were available.[14] A short holiday enjoying the 'natural beauty' of the hills, with their health, recreation, therapeutic and aesthetic associations, became attractive to a relatively small group of better-off educated and leisured tourists, to the puzzlement of locals. Surviving hotel visitors' books, diaries and newspaper accounts offer insights into such walking, merging as it did culture, landscape and philosophy.

The reasons for walking were various. Amato has suggested that many of such walkers were attempting to be 'true to themselves, nature and membership

---

13   J.K. Walton and P.R. McGloin, 'The tourist trade in Victorian Lakeland', *Northern History* 17 (1981), pp. 15–82.

14   *Times* (2 September 1854).

in greater spiritual communities'.[15] Anne Wallace sees them as trying to get in touch with a lost pastoral inheritance and inspiring and regenerating their lives.[16] Rebecca Solnit emphasises walking for pleasure as much as aesthetic or social meaning.[17]

A few hotel visitors' books include reference to walking routes. A surviving Salutation Hotel (Ambleside) visitors' book covers 1858 to 1874, and shows that already there were some foreign tourists, in 1868 for example, twenty-one from the U.S.A. Details included might cover place of origin, occasionally occupation, comments about the area, or references to activities undertaken. Books also survive from Wasdale Head (Wastwater), almost the geographical centre of the Lakes and thirteen miles by road to Drigg, on the Furness railway line, preserved because they contained details of rock-climbing routes. A visitors' book at Wasdale's Huntsman's Inn first mentions summit climbing in 1863, and there are later ones with climbing and walking references covering most dates from 1879 to 1904.[18] The books for the nearby small teetotal farmhouse, Row Farm, cover the period from 1876 to 1886.[19] One entry, of 3 June 1879, refers to as many as twenty-two people staying there at one time. The urban areas of Yorkshire, Lancashire and the south dominated guest addresses. Most entries came between March and September, with peaks during holiday periods. Many guests came from Oxford and Cambridge colleges, London's University College, Lincoln's Inn, hospitals, and various public school locations. There are occasional references to rucksacks or knapsacks, and to strenuous walking routes over the hills and passes. One entry of 21 June 1877 mentioned a round trip of thirty-two miles. Length of stay varied from a single day to a week or occasionally more.

Diaries, reminiscences and even letters to companies from customers using boots or walking equipment put further flesh on walking activity, and confirm that some individuals enjoyed a week or more of 'tramping', meeting other tourists en route from place to place. Others walked out from a base. The diary entries of Elisabeth Spence Watson show her regularly staying for three weeks or more at Grasmere, boating, sketching and walking long distances. In 1874, for example, she went to Borrowdale and Buttermere via Honister Crag, taking the coach back. She also tackled Harrison Stickle and Helvellyn with husband, sister and

---

15   J.A. Amato, *On Foot: History of Walking* (New York: New York University Press, 2004), p. 106.

16   A.W. Wallace, *Walking, Literature and English Culture* (Oxford: Clarendon, 1993).

17   R. Solnit, *Wanderlust: A History of Walking* (New York: Penguin, 2001).

18   For details see the Fell and Rock Club Climbing Archive, Cumbria Records Office, Kendal, Cumbria. See also M. Cocker, *Wasdale Climbing Book: A History of Early Climbing in the Lake District* (Glasgow: Ernest Press, 2006).

19   A Row Farm Guest House visitors' book for 1876–86 survives, and was part published in H. and M. Jackson, *Lakeland's Pioneer Rock Climbers* (Clapham: Dalesman, 1980).

children, hiring one or two ponies so that the youngest daughter could be carried.[20] Second-home owners also enjoyed walking. In 1898, for example, the Manchester brewery owner William Groves, his wife and children used their first summer at Holehird, near Windermere, to take a week's walking holiday across the central fells, walking first from Windermere to Buttermere, via Keswick, and then to Wasdale, Langdale and back.[21]

Rambling holidays were opened up to a wider audience through the initiative of T.A. Leonard, a Congregationalist minister from Colne in Lancashire. Under the auspices of the National Home Reading Union (N.H.R.U.), he pioneered simple and strenuous recreational and educational holidays in order to improve the social and moral conditions of his flock, by offering reasonably priced accommodation and by promoting friendship and fellowship amid the beauty of the natural world. Following the success of the initial visit to Ambleside in 1891 and subsequent visits to Keswick, he founded the Co-operative Holidays Association (C.H.A.) in 1897. Leonard had no wish to arrange a 'particular jolly type of holiday', but rather wanted, as Keith Hanley and John K. Walton argue, to encourage 'physical and moral renewal through strenuous walking in rugged scenic surroundings, proximity to nature and therefore Nature's God, and serious reading and discussion in the evenings'. However, as the C.H.A.'s holidays became physically less arduous and there was a gradual retreat from simplicity, Leonard left in 1913 to found the Holiday Fellowship (H.F.) in a renewed effort to establish holidays that would be genuinely working-class in appeal and composition, and the H.F. took over the C.H.A.'s 'Spartan' centre at Newlands.

Beyond climbing and walking we have limited statistical data on those activities related to informal, non-competitive incidental holiday sports, though hotel advertisements suggest that sporting activity was sought by at least some guests. In 1914, for example, some Lake District hotel and guesthouse advertisements in the Ward Lock *English Lake District* guide imply that sporting participation might be an attraction to visitors. More than 30 per cent of advertisements mentioned climbing, 23 per cent golf, and 15 per cent rambling, angling, boating, bowling and tennis. Lake District hotels increasingly provided lawn tennis facilities for middle-class guests. At Grasmere, for example, the Prince of Wales Lake Hotel and Cowperthwaite's Rothay Hotel offered facilities as early as the 1890s. Fishing was popular with some, and ad hoc fishing on the lake was even taken up sometimes by excursionists, as when in 1881 a rail trip from Blackpool arrived and excursionists discovered it was the first day of the season for perch fishing.[22] Somewhat surprisingly, despite golf's middle-class cachet, courses were few in the Lakes. Ambleside's small golf course, half way up Loughrigg, was more famous

---

20   E.S. Watson, 'Reminiscences and Family Chronicles', at http://web.ukonline. co.uk/benjaminbeck/esw2.htm (accessed 20 January 2011).

21   I. Jones, *The House of Hird: The Story of a Windermere Mansion* (Milnthorpe: privately printed, 2002), p. 80.

22   *Westmorland Gazette* (18 June 1881).

for its views than the quality of its fairways. Windermere's golf course, founded in 1891, also claimed breathtaking scenery.

As cycling slowly extended its middle-class appeal cycling tourists appeared. By 1897 the Cyclists' Touring Club (C.T.C) had 44,491 members. The pneumatic-tyred safety bicycle had become widely available, and together with better maps encouraged cycling tours of the Lake District. In 1900 Messrs Bartholomew of Edinburgh issued regional maps for cyclists and Cruchley's produced a *Westmorland & Lake District for Cyclists, Tourists. etc.* By *c.*1904 the Keswick Hotel marketed itself as the 'Headquarters of the Cyclists' Touring Club'.[23] Even United States President Woodrow Wilson came to cycle. He took long summer holidays in the Lake District on four occasions between 1896 and 1908, renting a cottage, and writing fascinating letters home describing his cycling rides.[24] Cheaper camping holidays were also becoming more popular for what Don Schumacher has called 'soft' sport tourism adventure, sometimes providing a basis for 'harder' sport activity.[25] In 1880, for example, at Whitsuntide, a 'canoeing party' were camping near Low Wray Farm near Windermere.

Tourist *spectatorship* at the far more rarely held formal sporting events was popular even at the start of the nineteenth century. Sailing and rowing regattas on the Lakes started early: at Bassenthwaite in 1780, at Windermere in 1796. From then onwards Windermere's annual midweek regatta became a major tourist attraction, usually held in August when wealthy visitors were plentiful, and in 1805 London's *Times* reported the 'forthcoming' date for Windermere's 'celebrated regatta'. As elsewhere, regattas were commercial ventures organised by local innkeepers, sometimes with associated 'ordinaries' and balls. Landlords acted as Secretaries, collected subscriptions for prize money, appointed stewards, organised handbill distribution and press advertising, and benefited financially from the sale of food and drink. Edward Hart of the Ferry Hotel, for example, was Secretary in 1857 when Charles Dickens visited. Dickens showed little interest in the regatta, which was something 'to please the ladies'. He enjoyed the local form of pole leaping which he described as 'one of the most graceful treats at the ferry', noting that some visitors entered sports events ('university men and the like' who 'soon found themselves outclassed'). But he was most interested in the wrestling and wrestlers.[26] He soon learned that because of the frequency of such summer events, there was 'a distinct race of professionals who live by the exercise of their thews and sinews', and that there was an element of fight fixing, which he tactfully called 'various little arrangements' made beforehand.

---

23 Anon., *The Concise Series of Guides No. 1: the English Lake District* (London: Geo Phillips and Sons, *c.*1904).

24 A. Wilson, *An American President's Love Affair with the English Lake District* (Morecambe: Lakeland Press Agency, 1999).

25 D. Schumacher, 'Foreword', in Hudson, *Sport and Adventure Tourism*, p. xv.

26 C. Dickens, 'Feats at the Ferry', *Household Words* (6 February 1858), p. 176.

Regattas offered amusement and the opportunity for display to the local gentry and social elite, especially colourful, sociable characters such as Mr. Curwen of Belle Isle, John Bolton of Storrs Hall (whose house parties included politicians and the titled, their wives and families) or Mr. (later Professor) Christopher Wilson of Elleray, the son of a wealthy Paisley manufacturer. Stewards soon included more prestigious industrialists and merchants from Liverpool, Bolton and Manchester with houses in the Lakes.

Such individuals attracted deference from socially aspirant visitors, though competitors occasionally shocked spectator sensibilities. A visiting governess in 1810 noted that after the regatta there 'was a 'footrace' between four men, running half-naked, and noted that 'the ladies, disgusted, every one left the ground; there were many of Fashion and of rank … and there was a much greater number assembled on that day, than there was at the preceding Regatta; of elegant, well-bred women I mean'.[27]

In 1824 over 50 'barges' were employed in a spectacular lake procession, led by Wilson, which included in an illustrious company Sir Walter Scott, William Wordsworth and Robert Southey, as well as the Foreign Secretary Canning, who spent a month at Storrs. Every inn was 'crowded with tourists'.[28] In early September 1833 London's *Morning Post* provided vivid details of the Windermere and Ullswater Regattas. From Ambleside's Low Wood hotel, the 'nobility and gentry who were sojourning in the Lakes' watched not just well-contested rowing competitions but also 'traditional' local events, including what was 'called a trail hunt, a favourite diversion of the yeomanry who live on the mountains and keep a few fox hounds', won by a dog brought from distant Cockermouth, and 'the chief point of attraction', the wrestling, which was 'very superior'. At Ullswater Regatta the Earl and Countess of Lonsdale were amongst the 'company' to watch the aquatic sports, as well as another hound trail, horse racing 'of a very inferior description', and the wrestling which 'wound up the festivities [though] the sport was but moderate'.[29]

Windermere's regatta was occasionally held twice annually, but sometimes missed altogether in some years in the 1840s and 50s. It was highly dependent on landlords' competence in generating funding, and bad weather was usually financially disastrous since revenue was tied to accommodation and drink sales. As at seaside resorts, local boatmen and others competed for prizes, variously in carriage boats, fishermen's boats, pleasure boats, skiffs and wherries. Other 'amusements' such as footraces, hound racing, pole leaping and (most popular by far) wrestling also featured.

Unfortunately, descriptions of visitor numbers and their places of origin are imprecise, only implicitly quantitative. The *Morning Post* claimed on 31 August

---

27	J.J. Bagley and E. Hall (eds), *Journal of a Governess, 1807–1825* (Newton Abbot: David and Charles, 1969).

28	*Morning Post* (20 August 1825).

29	*Morning Post* (11 September 1833).

1808 that 'the banks of the lake were crowded with spectators from all parts of the country'. This could have been using the word 'country', as in foxhunting, simply to mean the local area, or be a journalistic exaggeration.[30] In 1810, 'many strangers honoured the meeting'.[31] In 1833, the regatta was 'well attended' by 'the Nobility and Gentry who have been sojourning at the Lakes'.[32]

From the 1850s railway companies could exploit the potential of such events for tourism. In 1858 special trains were run from Liverpool Lime Street to Windermere Regatta, leaving at 6.15 a.m., though with tickets priced at five and ten shillings they were only for better-off travellers.[33] But by the 1860s regattas were becoming unfashionable. When the landlord of the newly rebuilt Ferry Hotel made an unsuccessful attempt at reinvigoration in 1879, many excursionists from Bradford, Leeds and Huddersfield chose to visit Ambleside instead. The 'Aquatic and Athletic Fete' attracted Lord and Lady Decies and a few other notables, but only perhaps a thousand paid for entrance to watch the wrestling, cycling and athletic sports, though the event attracted a few semi-professional participants from Newcastle, Liverpool and elsewhere, as well as local competitors.

Regattas were occasionally held on other lakes. Derwentwater's regatta was sometimes held in October to extend the tourist season, and its boat races, wrestling and horse races in 1810 reportedly attracted a 'vast concourse of people', though how many were visitors remained unclear.[34] By the 1870s it was held in August. The Ullswater Regatta of 1833, which the Earl and Countess of Lonsdale attended, had boat races, but also foot races, wrestling, a hound trail and low quality horse races.

From the 1860s, the Lakes, especially Windermere, offered scope for the technicalities, sophistication and conspicuous consumption of more regular yacht racing. Yachting provided a hobby for rich and privileged local residents, second home owners and visitors, allowing them to parade before and excel in the eyes of their own social group, and to impress the lower social orders whilst remaining apart, secluded on their boats or in exclusive areas of hotels and residences.[35] Earlier, yacht races had been rare though Mr. Wilson who 'had a great passion for boating', had maintained a small Windermere fleet.[36] More modern yachts had sometimes been imported even before the arrival of the railway. In 1831, for example, a Captain

---

30 *Lancaster Gazette* (5 August 1809).

31 *Lancaster Gazette* (11 August 1810).

32 *Morning Post* (11 September 1833).

33 *Lancaster Gazette* (24 July 1858).

34 *Lancaster Gazette* (13 October 1810).

35 O.M. Westall, 'The retreat to Arcadia: Windermere as a select residential resort in the late nineteenth century', in Westall (ed.), *Windermere in the Nineteenth Century*, pp. 34–48.

36 'Memoirs of Professor Wilson', *Times* (22 October 1862).

White brought up a London-built boat partly by road.[37] But by 1853 Mr. Greville's boat Truant was brought by rail to Windermere Station. Its race attracted both 'the elite and flower of Windermere banks and many others from a distance'.[38]

From 1857 relatively rich locals and second homeowners raced for a Challenge Cup alongside other races for cups either donated or subscribed for. They saw such sport as a way of demonstrating wealth and prestige, and took the lead in forming the Windermere Sailing Club in 1860, introducing an initial size limitation of 22 feet in length and handicaps, and in 1904 a 17-foot 'second class' category.[39] In 1887 the Club acquired a Royal Warrant, a mark of social approval, and Royal Windermere gained even more social capital when it became a member of the elite Yacht Racing Association in 1895.

Membership of the Club was at least as much a mark of social approval as of interest in yachting, part of the attempt by this social elite to create their 'own retreat to Arcadia as well as their own patterns of amusement'.[40] A sense of Windermere yacht racing's attractions to rich visitors as well as new settlers can be gained from the 1905 members list.[41] This showed that of the 120 members, 50 per cent came from addresses round the Lake, and in all 61 per cent from Cumbria. 24 per cent came from Lancashire and Cheshire, especially from Liverpool and Manchester, both of which had good rail links with Windermere; 7 per cent came from the London area.

Less wealthy visitors could merely watch regattas from the lakeside, or hire the many rowing boats available on the major lakes. Larger numbers of spectators enjoyed watching traditional and other Lakeland sports. Holiday periods such as Whitsuntide always brought in excursion traffic and many towns and Lakeland villages timed their galas and sports to bring in a little extra tourist revenue. The Friendly Societies at Bowness, for example, held a Whit Gala in the 1880s with professional and amateur versions of events such as wrestling, bicycling, flat racing and jumping, attracting over 3,000 spectators in 1880.[42]

But by the later nineteenth century it was the two most socio-culturally significant major events, Grasmere and the Pooley Bridge, Ullswater, Sports, which attracted the majority of tourist spectators. They were a form of regional social glue, providing cultural capital for the local elite and rich visitors in the grandstand. London's *Daily News* in 1875 described 'members of Parliament and clergymen, country gentlemen and dons', and even the Prince of Abyssinia,

37   *Lancaster Gazette* (9 July 1831).

38   *Preston Guardian* (16 July 1853).

39   W.B. Forwood, *Windermere and the Royal Windermere Yacht Club* (Kendal: Atkinson & Pollitt, 1905); B. Hall, *The Royal Windermere Yacht Club 1860–1960* (Altrincham: John Sherratt & Son, 1960).

40   J.D. Marshall and J.K. Walton, *The Lake Counties from 1830 to the Mid-Twentieth Century* (Manchester: Manchester University Press, 1981), p. 115.

41   Forwood, *Windermere and the Royal Windermere Yacht Club*.

42   *Westmorland Gazette* (22 May 1880).

there to see 'the sports for which the district is famed'.[43] While other events varied, local-style wrestling, hound trailing and a fell race were always central. In 1870 Grasmere had a fell race for local 'guides', wrestling for heavy-weights and light weights, a mile race, high leap, pole leaping, a boys' race, swimming race and hound trail. Other years included tug of war, sack race, long jump, boat race, 200 yards, 100 yards, local and open events, professional and amateur events.

Both events attracted substantial crowds, depending on the weather, with estimated numbers at times as high as ten thousand or more. Usually we know little of how far either participants or spectators travelled to attend, or why. Neither site was close to a railway, so competitors had to walk, have a horse, pay for a coach place or obtain a lift. The better off variously used coaches, wagonettes, charabancs, gigs or dogcarts. For Grasmere, the nearest railway station was Windermere, nine miles away, though lake boats could be taken to Wateredge at Ambleside. Pooley Bridge spectators' station was a little nearer, at Penrith.

A fairly loose surrogate measure of visitor attendance can be obtained from the later nineteenth and early twentieth century newspaper reports listing high-status attendees, since they provided place of residence. Their weakness is the unknown basis of their compilation. They may well only include those in the grandstand that the reporter managed to speak to, recognised or obtained from pre-bookings listed by the committee. Quite often the term 'and party' was used. This might at times simply mean an extended family, but also included disparate groups from hotels, as in 'party from Low Wood Hotel'. One attendance list of 1900 included the phrase no fewer than 65 times. A majority of addresses given were always within the modern National Park or wider Cumbria. Even so a substantial minority were visitors, and other addresses were hotels or rented accommodation, implying further visitor presence.

Grasmere had more high-status spectators. By 1897 the *Times* described it as occupying 'the foremost position amongst athletic gatherings in the north of England', drawing 'lord lieutenants and earls, Bishops and poets of high degree ... in well-horsed equipages', thanks in part to Grasmere's association with 'Wordsworth and his literary *confreres*'.[44] Five titled individuals attended in 1900, and six in 1890, as well as Doctors (five and two), clergymen (12 and five), and military officers (16 and 13). The vast majority of such individuals were relatively wealthy local day-trippers, coming predominantly from within the National Park, or from the rest of Cumbria, though as the table below indicates, a few travelled further or in a few cases were seemingly summer visitors (Table 9.1).

---

43   *Daily News* (23 August 1875).
44   *Times* (28 August 1897).

Table 9.1   Places of origin of listed attendees: Grasmere 1890 and 1900

From	1890	1900
Future Lake District National Park	28	53
Rest of Westmorland and Furness	13	15
Rest of Cumberland	9	5
Northumberland and Durham	0	2
Scotland	0	1
Lancashire and Cheshire	7	13
Yorkshire	7	6
Midlands	3	1
London and South	4	3
Overseas	2	1

We can also gain some sense of how far the more working-class *competitors* travelled since they were often listed by their address, providing some measure of distance travelled, though not of whether an overnight stay was required. With the advent of handicap footraces, professional runners visited Lake events. Tom Burrows from Rochdale, 95 miles away, won the 200 yards race each year at Grasmere from 1895 to 1899, but was beaten by an Edinburgh runner in 1900.

Given the exigencies of space, a useful case study of travelling distances is provided by data relating to Cumberland and Westmorland wrestling. In 1873 only 17 per cent of what were described as 'principal wrestlers' in that style came from the future National Park, a further 59 per cent from the rest of Cumberland, 6 per cent from the rest of Westmorland and south Cumbria, and 18 per cent spread across Durham, Northumberland, Lancashire and southern Scotland.[45] These elite wrestlers travelled regularly to the principal wrestling meets, from Morpeth to Carlisle, and from West Cumberland to Liverpool, and such distances would have necessitated overnight stays.

Some slept in barns or in friends' houses, but many walked long distances there and back. In the 1870s almost all but the elite of entrants even at more prestigious wrestling events made a return journey of less than fifty miles, based on road distance measurements. This could be walked, especially taking shorter cross-country routes. By the 1880s and 90s, samples taken from Grasmere entries suggest that an increasing proportion of competitors travelled long distances. In 1900, of 72 individual entries to all adult wrestling events, 57 per cent travelled up to fifty miles in total, 35 per cent more up to a hundred miles, and the rest even further. By 1913, with better cycles and some motorised transport appearing, entries and travelling distances were both up. Of the 117 entries to the middleweight

45   I.T. Gate, *Great Book of Wrestling References Giving the Last Two, Three and Four Standers of about 2000 Different Prizes* (Carlisle: Steel Bros, 1874).

competition, at least 54 per cent travelled between fifty-one and 100 miles, 6 per cent even further, and the origins of 7 per cent were not identifiable. The majority of entries at other organised sporting events at Grasmere, however, came from nearby though for more modern sports there were occasional distant entries by travelling semi-professionals.

Sports Tourism 1914–51

Between the wars, transport and social changes impacted on Lakes sports tourism. Though rail travel was still important, and many better-off visitors still stayed for a couple of weeks or more, for more affluent families motorcar touring holidays were becoming more possible, and provided more flexibility, whilst bus services also penetrated the Lakes. Motorcars were allowed round Grasmere's sports ring for the first time in 1920, when a record crowd of 15,000 attended. Despite the impact of the Depression, more people received annual paid holidays in the 1930s: in 1936, an estimated one and a half million workers, and by early 1938 almost four and a half million.[46] This increased numbers of British visitors. Americans also now visited the Lakes in larger numbers.

Though here again we lack detailed figures, it is clear that the increased interest in hiking from the 1920s onwards, associated particularly with the young, had its impact on the Lakes. In June 1934 the London, Midland and Scottish Railway was actively marketing special ramblers' train journeys from Lancashire towns to the Lake District.[47] The first arguments for making the Lake District a National Park also appeared. Vaughan Cornish, in 1930, could rhetorically celebrate Lakeland as 'the National Park of Great Britain', unrivalled in its combination of scenic beauty in a compact area, with a well-defined boundary, commenting that 'a strong walker can explore the whole region from a single centre'.[48] More rambling clubs were formed, usually with a lower middle-class or skilled worker membership, as walking holidays became more popular.

In response to demand, and in turn stimulating it, specialised accommodation in the Lakes emerged catering for hiking and similar informal sporting activities. Members of the men-only Wayfarers Club, interested in mountaineering, walking, skiing and cave exploration, visited the Lakes regularly, and in 1929 they purchased a Langdale barn and converted it into a club hut. The C.H.A. and H.F. flourished between the wars with no fewer than seven centres established in the Lakes.[49] Although archive

---

46   The National Archives, Holidays with Pay Committee Report, 28 April 1938, LAB1/1, pp. 12–20.

47   *Westmorland Gazette* (16 June 1934).

48   V. Cornish, *National Parks and the Heritage of Scenery* (London: Sifton Praed, 1930), pp. 29–33.

49   R. Snape, 'The Co-operative Holidays Association and the cultural formation of countryside leisure practice', *Leisure Studies* 23 (2004), pp. 143–58; K. Hanley and

material in the form of reports, brochures, magazines and surveys is available for scrutiny, the contribution of the C.H.A. and H.F. to the history of holiday provision in the Lakes has yet to be examined in detail. However, a forthcoming study should shed some light on the role of these pioneering organisations in the outdoor pursuits phenomenon and its impact on the Lake District.[50]

The Youth Hostel Association was formed in 1930, modelled on the German Wanderfogel movement, and supported by the C.H.A., H.F., the British Youth Council and National Council of Social Service.[51] By 1935 its Grasmere and Keswick hostels were the most popular in the country, and by 1938 it had opened twenty-three hostels in the area. The Ramblers' Association was formed in 1935. Youth organisations began bringing youngsters to the Lakes. According to *The Times* 'a posse of Girl Guides from one of the many camps for boys and girls which one cannot but be glad to see in these parts had a perfect view' of the sheep dog trials and hound trails at Applethwaite, near Windermere in 1925.[52] By the early 1930s Elterwater had holiday chalets and caravan parking facilities. The first tented 'holiday camps' in the Lakes opened, providing cheap accommodation for hikers, and trading on locality. The Derwent Holiday Camp, for example, described itself as 'the best situated camp in England'.[53] By 1938 Westmoreland Rural District Council noted that 'the spending of holidays and weekends in caravans and under canvas was a custom that was growing'.[54]

Single-sex groups and young couples sometimes took cycling holidays in the Lakes, helped by the C.T.C.'s bed and breakfast guide. When Arnold Robinson and his wife took a one-week tandem trip in 1938 he met other tandem couples from Hull and Liverpool, and discovered that in Keswick, he could choose from a long accommodation guide, 'most of them regularly catering for cyclists'.[55]

We know very little of the extent to which visitors took part in less formal sport at local golf clubs or local parks, though a comparison of the number of 'visitors' using the facilities at Windermere Golf Club in 1913 (1,735 visitors) with those in those in 1921 (3,082) and 1922 (3,100) provides a general indication of what may have been a brief post-war boom. Unfortunately there is no indication how many 'visitors' were regional day excursionists rather than holidaymakers. A more general growth in

---

J.K. Walton, *Constructing Cultural Tourism: John Ruskin and the Tourist Gaze* (Bristol: Channel View, 2010), chapter 5.

50   D. Hope, 'Adventure holidays in the hills: "improving" holidays for working people *c.*1930–2000, with particular reference to the English Lake District', PhD thesis (in preparation), University of Cumbria.

51   M. Collett, 'A short history of the YHA', *Cumbria* 1, 1 (March/April 1947), p. 11.

52   'Fell life and sport', *Times* (11 August 1925).

53   *Burrow's Guide to the Lake District* (Cheltenham: Burrow, 1930).

54   *Westmorland Gazette* (8 January 1938).

55   R. Robinson, 'A tandem tour in 1936', *Cumbria* (September 1987), pp. 348–50.

interest is shown by the rise of overall men and ladies' membership from 264 in 1913 to 319 in 1920.[56]

Before 1914 the Lake District's pleasure towns had offered little in the way of recreational sports facilities, though from 1882 Keswick's Fitz Park had aimed to provide a 'pleasure ground and a place of recreation', specifically for inhabitants and visitors, with tennis courts and bowling greens that one Manx visitor described as 'the largest, cleanest and best rolled' he had seen.[57] Local authority tennis courts and bowling greens became more common after 1918. Keswick's Hope Park, founded in 1925, included various putting, crazy golf and pitch and putt activities. Ambleside's White Platts Recreation Ground opened in 1923, with tennis courts and putting green, and after its council purchase in 1928 added hard courts and a bowling green. Bowness developed the Glebe in similar fashion during the same decade. Hotels and guesthouse proprietors increasingly assumed that some prospective guests were interested in making use of these local sports facilities during their holidays. In 1930 angling, boating, climbing, golf and tennis were all mentioned by over 20 per cent of the 20 Lakes hotels advertising in the Ward Lock Lake District guide.

Sailing continued on the Lakes, but petrol-driven motorboats also began to appear. In 1923 the Ambleside Water Carnival introduced races for motorboats and hydroplanes, and thirty-four craft entered in 1924. Local interest in this more 'modern' sport encouraged the formation of a Windermere Motor Boat Racing Club in 1925, and women's membership was introduced in 1927, when a new clubhouse was built. Regular races were held during the summer, despite opposition from those residents who held very different views of 'suitable' lake leisure. Windermere U.D.C. was generally supportive, giving permission for races though soon restricting them to certain parts of the lake.[58] As with yachting, this drew in visitors, and press reports show owners coming from Manchester, Yorkshire and elsewhere. Windermere also became a venue for the Motor Marine Association Championship Cup. The Lakes became a venue for water-speed record attempts, allowed by the Windermere U.D.C. because it believed that they 'bring many people into the district'.[59] In 1928 Miss M. B. Carstairs' speed of 60 m.p.h. in her 'wonderful Saunders-Napier Motor Boat "Estelle II"' was covered by *Pathe News*.[60] The deaths of Sir Henry Segrave and his mechanic Victor Halliwell in June 1930, however, caused adverse publicity, and a temporary cessation, and though there were occasional further British attempts in the 1930s, by 1939 public interest had shifted to Sir Malcolm Campbell's endeavours on Ullswater with his boat Bluebird II.

---

56   Information provided by Jane Maddock, Windermere Golf Club.

57   *Isle of Man Times* (23 August 1884).

58   C. O'Neill, 'Visions of Lakeland: Tourism, Preservation and the Development of the Lake District 1919–1939', PhD thesis, Lancaster University, 2000, pp. 129–34.

59   *Westmorland Gazette* (3 May 1930 and 21 June 1930).

60   *Pathe Gazette* (28 July 1928).

The closest equivalents to sporting mega-events in the Lakes were still the August Pooley Bridge and Grasmere sports. Their distance from large centres of population largely limited attendance to visitors and locals with time and money to spare. Five of the six of the injured in a bus accident returning from Pooley Bridge to Moresby Parks (Whitehaven) in 1925, for example, were in their late teens.[61] Because such sporting attractions in the Lakes were of interest to visitors, regionally and nationally, as well as locals, they were regularly noted in the metropolitan press, helping to frame future visitors' perceptions and providing a discursive framework. National attention now focused on Grasmere. After 1918, *Topical Budget, Pathe* and other newsreel companies broadcast the sports under titles such as 'Grasmere Sports', 'Grasmere Games', 'The Dalesman's Derby', 'Brawn and Skill', 'Stalwarts Vie at Grasmere Games' or 'Sportsmen of the North', bringing these 'old English games' to the attention of the wider population. Even BBC radio featured a report of the day's events at Grasmere as early as 1938, though only in its northern regional programme.[62] Many commentaries made the tourist links explicit. In 1951 *British Paramount News* gave their newsreel the title 'Grasmere Games Entertain Lakes Holiday Makers'.[63] The newsreel companies also gave much coverage to lake sports, such as yachting and motorboat speed attempts.

The events changed little over time, though cycle racing, a new 'invented tradition', was introduced experimentally in 1931, and became a feature from 1938, when entries came largely from Carlisle and Penrith, as well as Birmingham, Wisbech and Cambridge, an indication that some competitors were travelling longer distances. Further indications that modern transport widened the competitor base can be seen in wrestling entries. In Grasmere's 1930 and 1934 heavyweight events, for example, at least 18 per cent of competitors had to make a round trip of over 100 miles.

Conclusion

This study demonstrates both the challenges and opportunities of a study of sports tourism in the past. On the one hand, it shows how difficult it is to put any detail on key questions such as how many visited or competed, over what time period, how often, or how far they travelled, and this forces a more cultural, broad brush approach. The lack of quantitative data forces creative use of surrogate measures such as visitors' books, attendance lists or wrestlers' home addresses. Clearly the methodologies and approaches of modern sports tourism are difficult to apply in the past setting. But this study, whilst not of the modern urban areas previously

---

61   *Times* (31 August 1925).
62   *Westmorland Gazette* (13 August 1938).
63   *British Paramount News* (27 August 1951).

examined, also shows that qualitative data such as guidebooks, newspapers, newsreels or diaries can and do shed interesting light on past sports tourism.

The Lakes drew in substantial numbers of tourists, largely from an educated, wealthier background, but apart from occasional sports team tours, very few visitors came purely to engage in modern sports, though some spent weeks enjoying less competitive activities such as fell-walking, hiking or cycling. Most perhaps enjoyed a little rowing or fishing on the lake, or putting, tennis, golf or bowls in the course of their visit, though if their visit coincided with a local sports event either on a neighbouring lake or one featuring supposedly 'traditional' Lake District sports, they might well attend. Wealthier visitor competitors were drawn to elite events such as yacht or motor racing. For competitors, rises in living standards and better transport encouraged more working-class competitors to travel some distance to compete in sports such as wrestling or fell running.

This made Lakeland quite distinctive, throwing up patterns of sports tourism very different to those found in urban areas such as London or Manchester, with their globally famous soccer sides, their international venues and regular hosting of events. How distinctive, however, given the lack of published research on comparative work on areas such as Snowdonia or the Cairngorms, with their scope for 'adventure' tourism and their variations in social and cultural change, is yet to be determined, though both areas might well show up interesting parallels.

PART III
Lake District Tourism Case Studies

Chapter 10

Claife Station and the Picturesque in the Lakes

Sarah Rutherford

Introduction[1]

The Picturesque landscape aesthetic which developed in Britain from the mid-eighteenth century was particularly associated with the English Lake District and its tourism.[2] Claife Station (a station is defined in Picturesque terms as a viewing point) was a seminal element in the earliest days of tourism in the Lakes, designed with a single purpose – to facilitate and enhance the viewing of the dramatic scenery. Dozens of natural viewpoints were identified around the Lake District forming a now largely forgotten system of stations. There were five such stations around Windermere first described in the pioneering guide book to the Lake District published by Fr Thomas West in 1778.[3] Of these Claife was the most important and sophisticated, principally because it was the only one provided with a building, or pavilion, which served as a viewing platform for tourists. The station occupied a rock platform above the western shore of Windermere, at the point where the ferry arrived from Bowness, and was framed against a backdrop of yew and oak woodland.

---

1 Claife Station is owned by the National Trust. This chapter is based upon research and analysis commissioned by the Trust from the author and presented in a report: S. Rutherford, 'Claife Station, Cumbria: Framework Conservation Plan' (unpublished report for the National Trust, 2008). The author is indebted to the Trust for allowing her to draw upon this information.

2 For seminal information on the Picturesque in Britain see M. Andrews, *The Search for the Picturesque. Landscape Aesthetics and Tourism in Britain, 1760–1800* (Stanford: Stanford University, 1989). An extensive compendium of primary sources is given in The Helm Information Literary Sources and Documents Series: M. Andrews (ed.) *The Picturesque Literary Sources and Documents* (Robertsbridge: Helm, 1994). Specific works on the Picturesque in the Lake District include P. Bicknell, *The Picturesque Scenery of the Lake District 1752–1855: A Bibliographical Study* (Winchester: St Pauls Bibliographies, 1990) and on early tourist reactions to the Lake District: N. Nicholson, *The Lakers. The Adventures of the First Tourists* (London: Hale, 1955).

3 T. West, *A Guide to the Lakes: dedicated to the lovers of landscape studies, and to all who have visited, or intend to visit the Lakes in Cumberland, Westmorland and Lancashire. By the author of the Antiquities of Furness* (London: Richardson & Urquhart and Kendal: W. Pennington, 1778).

Although now a ruin (Figure 10.1), the once flamboyant viewing pavilion became an eye-catcher from the lake and its rugged and landscaped surrounds. Built in the 1790s, the two-storey pavilion, in the form of a free-standing octagonal tower, was designed, probably in the Neo-Classical style, by the renowned and prolific northern architect John Carr (1723–1807). Carr went on to design extensions to John Plaw's extraordinary cylindrical villa on nearby Belle Isle for John and Isabella Curwen (see Menuge, this volume) and in the 1800s it was John Curwen who, having acquired Claife, modified Carr's pavilion in the Gothic style with the addition of crenellated screen walls and a bay window. This window in the first-floor drawing room was jewelled with coloured glass borders, each colour intended to evoke a particular season or weather phenomenon in the surrounding scenery. Sensory experiences were enhanced by the provision of an Aeolian harp.

Figure 10.1   Claife Station's ruined pavilion awaiting conservation by the National Trust
Source: © Jason Wood.

As a tourist magnet of such an unusual and exciting form, the station became renowned and its cultural associations are numerous. The pavilion and its scenic effects were described by various literary and other visitors, most notably two lions of the emerging Romantic movement: William Wordsworth and Robert Southey. Wordsworth's *Guide to the Lakes*, published in five editions during his lifetime, from 1810 to 1835, sets the station in its landscape and tourism context,[4] Southey details the varied sensory experience of the interior with its coloured glass and Aeolian harp.[5] Artistically the station was illustrated, both by amateurs and professionals, usually as the backdrop to an event on the lake, and often with considerable license. De Loutherbourg presented the bare rock in 1785 before the pavilion was built as a breathtaking outcrop of sublime ruggedness above storm-wracked waters.[6] John 'Warwick' Smith's view of 1792 shows a more prosaic scene with the steep rock face overlooking the lake and ferry activities. Less accomplished later artists show the Gothicised pavilion imparting a plausible medievality to the scene which was otherwise absent.

Within the Lake District Picturesque context, this chapter will examine how Claife Station was enthusiastically embraced as part of the essential tourist experience, both aesthetic and artistic, in the late eighteenth and nineteenth centuries, becoming one of the most important cultural symbols of this pursuit.

The Picturesque Movement

Picturesque theory developed during the eighteenth century. The Picturesque was, together with the sublime and beautiful, the third element of a wider aesthetic movement based on an appreciation of scenery and a range of prescribed and acceptable emotions it engendered in the viewer. The word Picturesque derived from the French and Italian terms for 'painterly style', and was used in England in the early eighteenth century by the poet Alexander Pope in describing Homer's prose. It originated from Italian paintings of Classical scenes by seventeenth-century artists including the three masters, Salvator Rosa, Gaspard Dughet (who adopted the name Poussin), and Claude Lorrain. The range of emotions evoked was codified, its vocabulary including terms imbued with specific meanings such as beauty, horror, sublime and immensity.

Eighteenth-century English aristocrats who undertook the educational Grand Tour to the Continent encountered a range of landscapes on the journey considered to be picturesque. These included, in general terms, the terrors of the Alpine and Apennine passes and mountains, evoked by Salvator Rosa, and the more pastoral and Classical scenes in the Roman Campagna evoked by Claude Lorraine. These

---

4   W. Wordsworth, *Guide through the District of the Lakes .*, 5th edn (Kendal: Hudson & Nicholson and London: Longman & Co, 1835). See also the illustrated version of the 1835 edition: W. Wordsworth, *The Illustrated Wordsworth's Guide to the Lakes*, ed. P. Bicknell (Exeter: Webb & Bower, 1984).

5   M. Alvarez Espriella, pseud. [i.e. Robert Southey], *Letters from England: by Don Manuel Alvarez Espriella. Translated from the Spanish,* vol. 2 (London: Longman & Co., 1807).

6   The painting, *Belle Isle in a Storm,* is at the Abbot Hall Art Gallery, Kendal. The pair to it, *Belle Isle in a Calm,* dated 1796, shows calm scenery of the lake.

tourists' reactions to the Alpine scenery changed as the century progressed. At the beginning of the eighteenth-century travellers were frightened by wild scenery. By the century's end this fear had transformed into a positive liking for 'Salvator Rosa and Sublimity'. Travellers sought ever-wilder places and the writers of tourist guides responded to their new visual appetite. The reaction to the passage of the Alps gradually changed from a genuinely terrifying experience to one which induced awe and fear at the time but which could be recalled at home with excitement and youthful pride at the dangers overcome.

The British Picturesque was defined by its chief arbiter the Revd William Gilpin in his *Essay on Prints* ..., published in 1768, as 'expressive of that peculiar beauty which is agreeable in a picture'.[7] He did much to shift the emphasis of the term Picturesque from pictures to the landscape with a series of British guidebooks published between 1782 and 1809. These books helped to start a British equivalent of the European Grand Tour even before the Napoleonic Wars, with tourists descending upon the countryside, sketchbooks in hand, eager to experience this picturesque beauty and the emotions it aroused.

In Picturesque tourism two key interrelated paradoxes have been identified by Andrews. Firstly, although the viewer wishes to appreciate Nature untouched by man he cannot resist the impulse, if only in the imagination, to 'improve'. Secondly, it is the native scenery which is appreciated, but this is done by invoking idealised foreign models such as Classical Greece and Rome.[8] The first is more obviously relevant to Claife station, for the natural scene was 'improved' in the 1790s with the erection of the pavilion surrounded by a small pleasure ground. The coloured glass bordering the pavilion windows 'improved' the scene still further in the viewer's imagination and the Aeolian harp contributed further sensory stimulation. In terms of the second, the pavilion may have been constructed as an Athenian Tower of the Winds, evoking Classical associations, but if so it was quickly divested of this foreign association and turned into a firmly native Gothic fortification.

The Picturesque and Tourism in the Lakes

Upland wastes were shunned as barren and unattractive when Celia Fiennes first published an account of a journey through the Lakes in 1698. She commented with feeling of Windermere: 'as I walked down at this place I was walled on both sides by those inaccessible high rocky barren hills which hangs over ones head in

---

7   W. Gilpin, *An Essay on Prints: containing remarks upon the principles of picturesque beauty, the different kinds of prints, and the characters of the most noted masters, illustrated by criticisms upon particular pieces to which are added some cautions that may be useful in collecting prints* (London: J. Robson, 1768). Originally published anonymously.

8   Andrews, *The Search for the Picturesque*, p. 3.

some places and appear very terrible'.[9] Defoe concurred and thought the country in Westmorland 'the wildest, most barren and frightful of any I have passed over in England or even in Wales itself'.[10] Even by 1750 the Lake District remained largely a desert for people of taste; but perceptions were about to change radically. The first visitors to the Lakes who came specifically to enjoy the scenery arrived in the 1750s, predating the interest of the Romantic poets by some half a century. From the 1770s the lake scenery of Cumberland and Westmorland became a serious challenge to the aesthetic supremacy of the European Grand Tour, particularly with the inaccessibility of Europe during the political upheavals between the 1790s and 1815. Even after this hiatus in foreign travel the area remained a popular destination with the most highly regarded scenes described again and again in the numerous guide books.

By the 1790s 'a *Rage for the Lakes*' had sprung up, 'in which we travel to them, we row upon them, we write about them.'[11] The feeling for these sublime landscapes shifted so far and so fast that by 1810, the date of the first edition of his *Guide to the Lakes*, Wordsworth was noting that:

> persons of pure taste throughout the whole island ... by their visits (often repeated) to the Lakes in the North of England, testify that they deem the district a sort of national property, in which every man has a right and interest who has an eye to perceive and a heart to enjoy.[12]

The great improvement in the roads in the late eighteenth century encouraged travellers to visit. West attributed the increasing popularity of the Lake District to tourists to the improvement of the roads since Gray visited in 1765 and Pennant in 1772.[13] The familiarity engendered made visitors more complacent about such formidable scenery whose dramatic and dark nature was to attract the followers of the sublime: a combination of pleasure mixed with fear, which gradually became popular during the eighteenth century. Initially the chief attraction was viewing the lake scenery in its combination of beauty, horror and immensity which united to thrill the emotions, rather than toiling to reach the higher fells and peaks, but the Romantics found these spectacular places inspiring and later popularised them as destinations.

---

9   C. Fiennes, *The Journeys of Celia Fiennes*, ed. C. Morris (London: Cresset Press, 1949), p. 196.

10   D. Defoe, *A Tour Through the Whole Island of Great Britain*, ed. P. Rogers (Harmondsworth: Penguin, 1971), p. 550.

11   H. L. Piozzi, 'Journey through the North of England and Part of Scotland, Wales, etc.' (1789), John Rylands Library Eng. MS 623, f.167, quoted in Andrews, *The Search for the Picturesque*, p. 153.

12   Wordsworth, *Guide through the District of the Lakes*, p. 309.

13   West, *A Guide to the Lakes*, p. 2.

The lake scenery on the Picturesque tour to Cumberland and Westmorland had extraordinary visual effects as later described so evocatively by the Romantic poets Wordsworth, Coleridge and Southey. In particular the lake as a great mirror fascinated those viewing it with Picturesque sensibilities. The scenery reflected in each of the great lakes was imbued with its own peculiar character associated with one of the three masters of landscape painting. Guide books gave the lead in opinions. William Hutchinson in his popular *Excursion to the Lakes*, published in 1774 judged that Poussin's paintings 'describe the nobleness of Hullswater' although sometimes Ullswater was awarded to Claude.[14] West, in 1778, awarded the 'delicate touches of Claude' to Coniston, the 'noble scenes of Poussin' to Windermere and the 'stupendous romantic ideas of Salvator Rosa' to Derwentwater.[15] These identifications offered a guide to the amateur painter and sketcher in how to approach the composition and atmosphere as he or she selected and sketched from the viewpoints. The tourist may well have read Gilpin on appreciating and sketching landscape. His books instructed a new and more receptive audience on how they should view the landscape which they could now more easily reach. His descriptions of picturesque scenery were followed by essays which set about translating these picturesque scenes into drawings and paintings, and which reached the intellectual heart of eighteenth-century tourism: landscape painting and the appreciation of nature.[16]

Picturesque tourism, with its paraphernalia and theories, was sufficiently well established by the end of the eighteenth century to be satirised. Rowlandson mocked it in Dr Syntax's *Tour ... in Search of the Picturesque*. Our hero not only tumbled into a lake while sketching a Gothic scene but was again soused when his nag Grizzle stumbled while he was sketching and not paying attention to her.[17] The attitudes were further ridiculed in a comic opera *The Lakers*, by James Plumtre in 1798.[18] Miss Beccabunga Veronique boasts of embellishing the scenery she paints to achieve picturesque veracity as

> if it is not like what it *is*, it is like what it ought to be. I have only made it
> picturesque ... I have only given the hills an Alpine form, and put some wood

---

14   W. Hutchinson, *Excursion to the Lakes, Aug. 1773* (London: J. Wilkie & W. Goldsmith, 1774).

15   Quoted in Andrews, *The Search for the Picturesque*, p. 159.

16   W. Gilpin, *Three Essays: On Picturesque Beauty; On Picturesque Travel; and On Sketching Landscape* (London: T. Cadell & W. Davies, 1808).

17   See J.J. Savory, *Thomas Rowlandson's Doctor Syntax Drawings: an Introduction and Guide for Collectors* (London: Cygnus Arts, 1997), pls. 13 & 18; also Hanley, this volume.

18   The very name by which Lake District tourists were known, the 'Lakers', had an underlying irony, since 'laking' or 'laiking' is the dialect word for 'playing' and more particularly for playing as children play.

where it is wanted, and omitted where it is not wanted; and who could put that sham church and that house into a picture?[19]

She expresses Gilpin's recommendation that the touristic connoisseur should mentally falsify (or as he would say, adapt) the scene before his eyes, setting to right Nature's deformities, and adding extra accessories where desirable. He qualified this practice by rules, however, that the artist should introduce nothing alien to the scene, with alterations kept in character with the country.

Thomas West and the Windermere Viewing Stations

The popularity of visits to the countryside in eighteenth-century Italy, particularly by the English aristocracy, created a demand for painted *vedute*, or views, to record places of interest on what became almost fixed tours, particularly in the Roman Campagna. Guide books were written for the Italian Tours with a list of places to visit such as Marmore Falls in Umbria where artists provided pictures rather like a photograph as a souvenir, and the guidebook sometimes included *vedute*.

This idea was apparently translated from Italy and codified precisely to the Lake District setting, starting with West's *Guide* of 1778. West was the first to record systematically the best places to view the lakes, identifying these as stations. His extensive description of Lake District scenes and views, and what to appreciate in them, was the most influential for fifty years or so, although other guide books were published and continued the trend. West's choice of stations was presumably based on those that were most well-known and in his opinion the most visually rewarding. There was no set number for each lake; for example, Coniston had three, Derwentwater had eight, Ullswater had four and Windermere five, the number being regardless of their relative sizes. West's promotion of these viewing stations was not original. Many had been identified by earlier visitors and in particular by Thomas Gray, Arthur Young, Dr John Brown (in a letter written 1753, published in 1767) and Dr John Dalton (in a poem published in 1754). Brown's letter is usually regarded as the first distinct evidence of a romantic and picturesque response to the scenery of the Lakes, but Gilpin in his *Dialogue ... at Stow* had also paid tribute to the region in 1748.[20]

In his *Guide*, West noted that the 'taste for landscape' had lately induced many to visit the Lakes who wished to contemplate the 'Alpine scenery, finished in nature's highest tints, the pastoral and rural landscape, exhibited in all their stiles, the soft, the rude, the romantic, and the sublime'. The intended audience was those who practised the 'noble art' of landscape painting as well as more casual

---

19   This must refer to Pocklington's Island on Derwentwater.

20   Andrews, *The Search for the Picturesque*, pp. 158–9; W. Gilpin, *A Dialogue upon the Gardens of the Right Honourable the Lord Viscount Cobham at Stow in Buckinghamshire* (Los Angeles: William Andrews Clark Memorial Library, 1976).

visitors. He intended to evoke in miniature, using 'nature on a reduced scale', the Continental Alpine region for those who intended to travel there afterwards, or implicitly to those who were never likely to reach the Alps. He set out the best sights for the traveller, and ways of viewing them. He indicated 'all the select stations, and points of view, noticed by those who have made the tour of the lakes … with remarks on the principal objects as they appear viewed from different stations'. It was important for the traveller to experience the 'agreeable surprise' on 'the first sight of scenes that surpass all description, and of objects which affect the mind of the spectator only in the highest degree'. [21]

Windermere, West proclaimed, was the 'finest lake in England'. His suggested approach from Esthwaite Water to the west through Sawrey opened up with first a long view south along its length as the road approached the western shore. His five Windermere stations were shown on Peter Crosthwaite's map of the lake (1783–94) and were confined to an area around the centre of the lake, easily reached by boat as a detour from the main route between Hawkshead and Ambleside (Figure 10.2). Claife Rock on the west side of Windermere was the first of West's stations. The Rock, originally part of the Claife parish common land, a tract of steeply sloping ground unsuitable for cultivation, was already popular as a panoramic viewing point due to its flat, elevated position close to the ferry. West's description of the panorama was characteristically detailed. From the plateau, two views commanded 'all the beauties of this magnificent lake'. The second and third stations were at the south and north ends of Belle Isle respectively, and the fourth was Rawlinson's Nab to the south, a peninsular rock swelling to a crown in the centre and covered with low woodland. The fifth station occupied a promontory above Bowness around Brant Fell at some distance from the shore and was unusual in being so elevated that the viewer looked down on most objects and took in the most expansive and panoramic view of the lake; viewing stations were generally preferred closer to the level of the water for artistic reasons. [22]

The time of day offered a variety of viewing opportunities. For example, West advised that the stations enjoyed the best views to the opposite sides of the valley at different times of day: from above Bowness the west side was best seen early in the morning when lit by the rising sun in the east; from Claife, the eastern, Bowness side of the valley was best seen later in the day when the sun had moved round to the west behind the western slopes.

Claife Pavilion

For the early Picturesque tourist Claife Station was merely a natural promontory from which to gain a prospect. De Loutherbourg's views of the Rock (1785–1786)

---

21   West, *A Guide to the Lakes*, pp. 2–3.

22   All five Windermere stations still survive and the views from them remain much as West described. Three are open to the public; the two on Belle Isle are not.

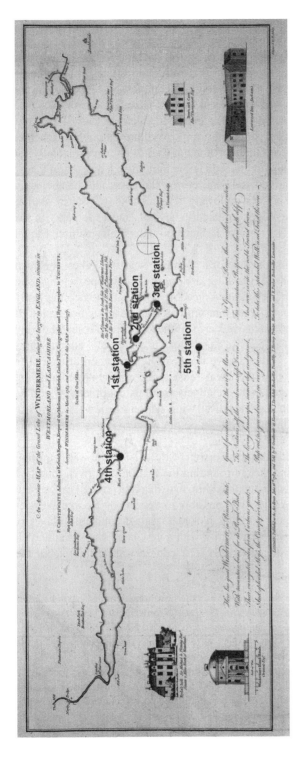

Figure 10.2 Peter Crosthwaite's map of Windermere of 1783–1794 showing West's five stations

Source: © Wordsworth Trust.

show it open and bleak. It was not yet subject to 'improvement'. That was to occur in the 1790s with the erection of an octagonal pavilion for the land owner, the Revd William Braithwaite. The date of its construction is unclear but it could have been begun as early as 1794. The building was complete by 1799, only a year before Braithwaite's death. The advertisement for its subsequent sale in 1800 (see below) notes that it was completed to 'a design by Mr Alderman Carr of York'.[23]

The form and architectural style of the Carr's pavilion are unclear, whether it was Neo-Classical or, as later, Gothic. Although Carr is known principally for his Neo-Classical designs he also produced Gothic buildings. However, at Claife a surviving fragment of the curved and cantilevered staircase, and the enclosing angled walls suggests that the original style may have been Neo-Classical, similar to an Athenian Temple of the Winds imitated in several landscape parks in the mid-late eighteenth century (for example, Shugborough, Staffordshire; Mount Stewart, County Down).

The 1800 advertisement for the sale of a 'Temple Near the Lakes' included surrounding 'pleasure grounds', indicating a designed landscape setting, and some twenty acres of good arable land and meadow. The Station was 'the justly admired TEMPLE' with a reputation as a tasteful erection of some pretension. It commanded the 'best and most extensive Views of the Lake and the neighbouring Mountains' being built 'at great expense', and those to whom this advert seems to have been directed would no doubt have heard of the reputation of Carr. The advertisement also assumed that the local reader, and more particularly the audience who might be interested in this relatively small parcel of land, would be educated enough to know of the concept of viewing stations, as it was referred to as 'Mr West's first and most favourite station on the western banks and near the centre of Windermere Lake'.

The Station was sold in 1801 to John Curwen of Belle Isle, who was acquiring much land in the area including that surrounding the Station.[24] Curwen modified Carr's octagonal building, enclosing it within two flanking crenellated screen walls with loopholes in the Gothic style, possibly to designs by George Webster of Kendal.[25] This more prominent building evoked a fragment of medieval origin, and enhanced the viewing experience from it with a new large bay window with coloured glass panes overlooking the water. The curtain wall at the rear of the

---

23   Revd William Braithwaite had a house at Satterhow, long demolished. For further detail see T.W. Thompson, *Wordsworth's Hawkshead* (Oxford: Oxford University Press, 1970) and H. Colvin, *Biographical Dictionary of British Architects, 1600–1840* (New Haven & London: Yale University Press, 1995), p. 255. The advertisement appeared in the *Cumberland Pacquet*, 2 (September 1800).

24   Curwen's acquisitions included in the 1780s the 'Great Boat' estate surrounding the Station as well as woodlands to the north: CRO/D/Cu Estate Plans 10. He was responsible for the extensive afforestation of Claife Heights and initiated the present wooded landscape to which the area owes much of its character.

25   Janet Martin, pers. comm.

station was extended to give more mass to the structure when viewed from the opposite shore and to contain additional outbuildings including a kitchen and storeroom. The building's presence and authority in the scenery were accentuated by whitewashing.

If real ecclesiastical or military remains with picturesque settings such as at Tintern, Rievaulx and Fountains Abbeys were unavailable, then faux Gothic features were resorted to, often within newly landscaped grounds. Claife's Gothic appearance was not intended to be a ruined folly but a building in good order. It may have represented the remains of a more extensive structure such as a local pele tower, perhaps as a reclaimed ruin, rebuilt and inhabited. The position, exposed high on a strategic promontory with a panoramic view of the lake, emphasised this defensive pseudo-association. It echoed in miniature the idea and style of Lyulph's Tower (*c.* 1780) on Ullswater, built shortly before the pavilion as a hunting box for the aristocratic Howards of Greystoke. This was the first picturesque, medievalising house in the area. Compared with that particular 'pleasure-house' (as Wordsworth referred to it in his *Guide to the Lakes*) Claife pavilion was diminutive and only designed to accommodate ephemeral visits for the purpose of viewing the scenery. From a distance both structures are credible in their deception as medieval fortifications. Southey in 1802, however, was not deceived and was critical of the style of Claife's 'castellated building ... standing upon a craggy point'. He found it unconvincing as it was built 'in a style so foolish, that, if any thing could mar the beauty of so beautiful a scene, it would be this ridiculous edifice.' However, once inside the building Southey found that 'this absurdity is not remembered ... and the spot is well chosen for a banqueting-house'.[26]

Wordsworth also referred to Claife pavilion as a 'pleasure-house', for which he had but faint praise. He regarded it as being 'happily situated' and 'well in its kind' but found that art had overreached herself as he still sighed 'for the coming of that day when Art, through every rank of society, shall be taught to have more reverence for Nature. This scene is, in its natural constitution, far too beautiful to require any exotic or obtrusive embellishments, either of planting or architecture'. Even the lodge-keeper who guided visitors to the Station contributed to the opening of the Picturesque scene: 'an aged female, inhabiting a pretty cottage within the inclosure surrounding the Station'.[27]

It certainly was purely a pleasure house, and one of some sophistication. A dining room and wine cellar occupied the ground floor, while the elegant first-

---

26 Alvarez Espriella, *Letters from England*: quoted in Andrews, *The Search for the Picturesque*, p. 166.

27 W. Wordsworth, *The Early Years, 1787–1805*, rev. C.L. Shaver [Early letters of William and Dorothy Wordsworth] (Oxford: Clarendon Press, 1967), p. 271. Claife pavilion and Lyulph's Tower were the only two buildings with this epithet in the first edition of Wordsworth's *Guide to the Lakes*, first published anonymously in J. Wilkinson, *Select Views in Cumberland, Westmoreland, and Lancashire* (London: Ackermann, 1810). See also W. Green, *The Tourist's New Guide* ... (Kendal: R. Lough & Co., 1819).

floor drawing room was reached by a stone spiral staircase at the rear, entrance side. The drawing room contained the famous three-sided bay window, described by Southey:

> The room was hung with prints, representing the finest similar landscapes in Great Britain and other countries, none of the representations exceeding in beauty the real prospect before us. The windows were bordered with coloured glass, by which you might either throw a yellow sunshine over the scene, or frost it, fantastically, tinge it with purple.[28]

It enjoyed two main views of the lake, the one towards Belle Isle being considered as

> equal in its kind to any other on the lakes, for it has every essential for a beautiful landscape; bold foreground, a fine transparent sheet of water, graced with islands, rich woods, and wavy mountains. It is an assemblage almost invariably grateful to the eye of the stranger, and all lovers of the beautiful in nature return to it with insatiable delight, but though so exquisite in nature, it is not easy of management in art.[29]

The pavilion and the magnificent views made the station a chief attraction of any holiday around Windermere and the patterns of colours playing around the interiors on sunny days would have been magical. Jonathon Otley described the effect of the coloured glasses and his perception of how the landscape was affected to evoke various seasons. The windows contained yellow glass to represent summer, orange for autumn, light green for spring, light blue for winter, a dark blue for moonlight and a lilac tint to give the impression of a thunder storm. Otley remarked that 'the view towards the north has every essential for a beautiful landscape; a bold foreground, a fine sheet of water graced with islands'.[30] James Henry Barber in 1838 concurred in some respects but not in others. He described the blue window with 'every object seen through a cold and wintry appearance' and saw green for spring, but saw red for summer and yellow for autumn, and 'through the purple you seem to behold a thunderstorm just ready to burst upon the lake'.[31]

The pavilion was open to the public and became a favourite resort of visitors, sometimes en masse and en fête. At the height of its fame during the 1830s and 1840s it was used for dinner dances held by the Curwen family. The drawing room

---

28   Alvarez Espriella, *Letters from England*: quoted in Andrews, *The Search for the Picturesque*, p. 166.

29   Green, *The Tourist's New Guide*.

30   J. Otley, *A Concise Description of the English Lakes, and Adjacent Mountains …* (London: Simpkin & Marshall, 1837), p. 4.

31   MS Journal of James Henry Barber: 'English Lakes Tour, 1838'. I am grateful to Diana Matthews, Barber's great, great granddaughter for allowing me to use this extract.

had a 'springy' floor. Music was provided by a stringed band. 'Winding walks' round the Station were lit up with Chinese lanterns and coloured lamps, to create a 'charming promenade'. The greatest novelty was crossing the lake by boat.[32] Opening hours according to a guide published in 1882 were 6am–9pm, May to October. Royalty graced its doors, and a visitor book recorded an entry made by Queen Adelaide when she visited in 1840.

The setting of the building was important, particularly the combination of artificial and natural landscape. Although the pavilion was enhanced by 'pleasure grounds' mentioned in the 1800 sale advertisement, these were subservient to the natural setting. Artists' views emphasise the Picturesque impact of the pavilion and the screen walls within a naturalistic setting. A painting by Mary Dixon (*c.*1810) and sketches by John Downman (1812) and George Webster (*c.*1850) show the building surrounded by trees largely obscuring the rock and lower storey so that the upper floor emerges in isolation from the rest of the building (Figures 10.3 and 10.4). At the foot of the rock were a crenellated wall and gateway screening a courtyard with a low lodge where the lodge keeper latterly had a tea garden and provided refreshment for visitors. It was here that visitors were first initiated to the experience after landing from Belle Isle. From the courtyard, a path to the Station curved around the slope in a gentle gradient cutting through the rock, walled on the outer side, to the top and the entrance front of the pavilion. This 'excellent road to the building … [was] graced on each hand by oak, ash and birch trees, springing from the sides and out of the fissures of picturesque rocks'. These trees were not just supplemented by commonly used ornamental evergreens but by 'an abundance of garden and field flowers, all filling the eye with a most pleasing assemblage of nature and art'. The breathtaking views were thus screened by the wall and the planting until seen from the interior of the pavilion.[33]

Other Viewing Pavilions in the Lakes and Elsewhere

The purpose-built viewing pavilion marked out Claife as the most sophisticated station in the Lakes. The pavilion was part of the fashion to create buildings from which to view dramatic scenery and its features and to act as incidents in journeying through the landscape. This had become well established in eighteenth-century parks, with pavilions and temples strategically situated to view the designed landscape and the setting beyond. Often the concealment of the views and their surprise revealing was an integral element of the design, employing architecture and planting, as at Claife.

---

32   M.M. Higginson, *Holidays in Lakeland … 1831–32* (*c.*1968).

33   The National Trust acquired the gatehouse, courtyard and lodge in December 2010, reuniting the pavilion with its approach. The long term aim is to reinstate the original access.

Figure 10.3 Mary Dixon's painting of Claife Station from the south, with Storrs Hall to the right, *c.* 1810

Source: © Wordsworth Trust, courtesy of Pamela Woof.

Figure 10.4   George Webster's sketch of Claife Station from the south, *c.*1850
Source: WDX 1315 Sketch Book of George Webster, architect of Eller How, Lindale.
© Cumbria Archive Centre, Kendal with kind permission of Rowland Hart Jackson.

The earliest purpose-built viewing pavilion in the Lakes was only ten miles away at Rydal Hall, Grasmere, for Sir Daniel Fleming, dating to 1668–69, a century or so before Gilpin and the Picturesque movement started to gain momentum. This remained a rare example for some time as buildings for viewing specifically natural scenery did not become popular until after the mid-eighteenth century. The pavilion or garden house at Rydal was built to frame and enjoy views of a waterfall and plunge pool. It was too small for anything else. The simple, stocky stone building had a door on the approach side from the garden. A large removable window on the opposite side gave a view of the waterfall, the plunge pool and the bridge above it. The visitor was led along a route to the garden house in such a way that the key view was concealed until the door was opened, revealing the waterfall framed by the window in the opposite wall. The scene from the window was painted as early as 1682 and commended by Thomas Gray in 1769 and by Gilpin in 1786, amongst many others. The scene was described in verse in 1794 by Wordsworth (who lived in the adjacent Rydal Mount from 1813 to 1850) in *An Evening Walk,* and the falls were painted by Joseph Wright of Derby in 1795. It was a *beau ideal* of romantic and picturesque scenery and drew many tourists; but because of its sunken position it was never a designated station.[34]

---

34   English Heritage, *Register of Parks and Gardens, Cumbria* (Swindon: English Heritage, 1999).

Other station buildings, more contemporary with Claife, formed part of landscape schemes related to country houses. West's second and third stations on Windermere, at the north and south ends of Belle Isle, were provided with viewing pavilions in the form of small summerhouses designed by Thomas White (*c.*1736–1811). The aforementioned Lyulph's Tower on Ullswater and Conishead Priory tower above Ulverston and the Lancaster coast were further examples. Beyond the Lake District early viewing pavilions for tourists of natural scenery include the two-storey view house at the falls of Cora Linn, Strathclyde (1708), and the Banqueting House, Hackfall, North Yorkshire (*c.*1750), built on the edge of a cliff overlooking a tumbling river. At Hackfall, the earliest sublime landscape in England, the vertiginous balcony cantilevered over the edge of a precipice to engender the Burkean sublime – the fear of falling combined with a vast horizon. From 1752 at Piercefield, near Chepstow on the Welsh border, Valentine Morris created ten principal viewpoints along a spectacular route above the River Wye, chosen 'with scrupulous attention to Picturesque prescriptions for organised prospects', including a Druid's Temple.[35]

The Hermitage at Dunkeld House, Tayside, en route to the Scottish Highlands, is a key precursor to the Claife pavilion, with notable similarities in its siting, form and elaborate design. It was built by the Second Duke of Atholl's nephew and heir, John, in about 1758. Like Claife it was sited on a rocky viewing point, reached via a route which prevented visitors from seeing the views, which were only revealed in a dramatic fashion through the windows of the main viewing room, embellished with coloured glass to 'improve' the scene. While the setting of The Hermitage was more immediately sublime and frightening than Claife, both used an element of surprise in the approach, concealing, as at Rydal Hall, the subject of the sensory experience until the very last moment. The Hermitage's red and green coloured glass was added to its windows between 1762 and 1783, intended to reproduce the atmospheric effects of the seventeenth-century artists. This was removed, possibly in response to criticism from cognoscenti such as Gilpin who visited in 1776, who thought it 'below the dignity of scenes like this'. Rebuilt and renamed Ossian's Hall in 1783 the interior was transformed into a stunning and bewildering kaleidoscope which evoked 'shock' and 'amazement'.[36] The most striking aspect of the new interior decoration was its lining with a proliferation of mirrors to create a 'glass drawing room' which made the spectator imagine that

---

35   Andrews, *The Search for the Picturesque*, p. 106.

36   The blind bard Ossian is the narrator and supposed author of a cycle of poems which in the 1760s the Scottish poet James Macpherson claimed to have translated from ancient sources in the Scots Gaelic. These became very popular and included the epic on the subject of the hero Fingal. Ossian was particularly associated with ancient Highland culture and in the late eighteenth century the so-called Ossianic Sublime. Ossian's Hall has recently been restored by the National Trust for Scotland.

the water was appearing from all angles.[37] The roar of water from the falls further befuddled the senses. By the late eighteenth century, the landscape along the River Bran was frequently visited as part of guided tours from Dunkeld and Ossian's Hall became one of the most visited stations on the Highland Tour. It evoked much spontaneous delight as intended, although Wordsworth's critical reaction echoed that of Gilpin's to the coloured scenery of its earlier incarnation: 'I am not aware that this condemnatory effusion was ever seen by the owner of the place. He might be disposed to pay little attention to it; but were it to prove otherwise I should be glad, for the whole exhibition is distressingly puerile'.[38]

Experiencing Scenery and Artistic Interpretation

Various devices were used to enhance the Picturesque experience which could be manipulated and intensified by a distinct set of artificial aids. These assisted the tourist to appreciate the scenery in various sensory ways, and evoke particular emotions. Claife Station is associated with several such devices, both directly via the pavilion architecture, and indirectly via its close association with West's *Guide*.

Claude and Gray glasses, along with the telescope and the Aeolian harp, were portable instruments common to the tourist's armoury. The Gray glass was a slightly convex mirror of about 4 inches across tinted with a black back foil, bound up like a pocket book. West recommended it as 'the landscape mirror will also furnish much amusement … Where the objects are great and near, it removes them to a due distance, and shows them in the soft colours of nature, and in the most regular perspective the eye can perceive or science demonstrate'.[39] Using this, the tourist could compose and reduce the large landscape into a neat, manageable scale, whether just to view it and respond emotionally, or to record it in watercolours. The convexity miniaturised the reflected landscape. Except in the foreground the details were largely lost, creating an idealised view. The Claude glass had transparent tinted filters in a variety of colours. Crosthwaite sold 'Gray's Landscape Glasses and Claude Lorrain's Do.' at his Museum in Keswick. A retail opportunity was not to be missed and these became popular devices for discriminating tourists.

This range of coloured tints evoked various seasons or types of weather and probably formed the model for the glass in the windows at Claife pavilion,

---

37 C. Dingwall, 'Gardens in the Wild', *Garden History* 22, 2 (1994), p. 144. This room of glass echoes the decoration in the earlier Gothic Temple at Bramham Park, West Yorkshire, based on a design by Batty Langley published in 1742. This too was mirrored (C. Gallagher, pers. comm., National Trust, September 2007).

38 W. Wordsworth, 'Memorials of a Tour in Scotland, 3: Effusion in the Pleasure-Ground on the Banks of the Bran, near Dunkeld', in *The Complete Poetical Works of William Wordsworth*, vol. 5 (1806–1815), p. 272.

39 West, *Guide to the Lakes*, p. 15.

which, in turn, provided an architectural alternative to the portable Claude glass. A similar palette suggested the various seasons and weather. Blue or grey glasses could suffuse a varied afternoon scene with moonlight or the cold colour of a winter day; the yellow or 'sunrise' glass, when used at noon, afforded a glowing dawn view, or represented autumn tints. Green could represent spring, and purple the effect of a threatening storm. The coloured glasses were also useful for introducing a tonal harmony into the scene, which was important for the painter brought up to respect the 'master tone', and could conveniently offer a near facsimile of such idealising effects.

Other sensory experiences were important, particularly sound. An Aeolian harp was kept at Claife (according to Southey), not surprisingly, as the instrument was popular in Germany and England with the Romantic movement of the late eighteenth and nineteenth centuries. A sound-box strung with several strings of varying thicknesses was placed in a window and gusts of wind across the strings produced an array of harmonic overtones. Poetry about the Aeolian harp began in England in the 1740s, with James Thomson who was fascinated by the instrument and pursued its association with the poetic mind, particularly in his 1748 *Castle of Indolence* praising 'Wild-warbling Nature all above the reach of Art'. J.M.W. Turner painted a Claudean scene entitled the *Thomson's Aeolian Harp* exhibited in 1809 at The Tate. Other sounds, on a more heroic scale, included cannon fired on pleasure boats out on Windermere and Ullswater creating a succession of echoes off the hillsides that tourists could thrill to, or sometimes the blowing of French horns.

Conclusion

Claife Station holds a key position in the history of tourism in the Lake District and is symbolic of the indigenous British Picturesque movement as it emerged and flourished from the mid-eighteenth century onwards. Most lake stations remained open viewing points without structures but Claife was unique in its architectural appearance and setting within its own station pleasure ground, and rare as a purpose-built structure. The pavilion is an elaboration of the simple idea of viewing nature in its raw form (but through cultivated eyes) on an open spot. It developed this concept by artificially directing the perception of the scenery to represent it in different seasons and weathers, by using an architectural framework and elaborate devices as the variously coloured glass to manipulate the senses. It also provided shelter for luckless tourists in the likely event of inclement weather. Its purpose and form derived from a long pedigree of designed landscape viewing buildings and points which was applied by the Picturesque movement cognoscenti to the appreciation of wider 'natural' landscapes. Other lakes had stations at specified points from which to view dramatic natural scenery but it was the purpose-built viewing pavilion crowning the rock that made Claife the most important station in the Lakes.

Chapter 11

Furness Abbey: A Century and a Half in the Tourists' Gaze, 1772–1923

Jason Wood

Furness Abbey has been a roofless ruin for 475 years; that is 65 years longer than it was a monastery. As such it has had a significant afterlife as the focus for romantic excursion and tourist curiosity. Despite its remoteness from the 'conventional' Lake District, and the difficulties posed to early travellers by the tidal crossings of Morecambe Bay, a visit to the abbey was considered an essential part of any trip to the Lakes for the discerning visitor from the late eighteenth century onwards. The observations of a veritable 'who's who' of Georgian and Victorian writers, and some curiously minded Americans, led to the abbey being acknowledged as a well-established aspect of the picturesque qualities of the Lake District. Consequently, and throughout the nineteenth century, the abbey became a site of considerable imaginative appeal to tourists, who came in increasing numbers following the arrival of the Furness Railway and the decision to build a station and hotel at the site. The period under discussion ends in 1923, as this was the year when ownership of the abbey, and coincidentally the railway station and hotel, changed hands.[1]

The abbey of St Mary of Furness is situated in the southern part of the Furness peninsula. Formerly known as Lancashire north-of-the-sands, the area's situation between the waters of Morecambe Bay and the inhospitable hills of Lakeland was one of singular isolation. Nevertheless, the abbey's location did make it vulnerable to Scottish raids from the north. The area was attacked in 1316 and six years later Robert the Bruce headed another raid. This time, however, the abbot of Furness, John Cockerham, begrudgingly entertained Bruce at the abbey. More unwanted guests arrived on 23 June 1537 when Robert Southwell, receiver of the Court of Augmentations, along with three other of Henry VIII's commissioners, obtained the deed surrendering the abbey to the king and set about surveying and dismembering the monastic buildings and disposing of the abbey's estate.[2]

---

1 Key references for the abbey's aesthetic qualities and tourism appeal are C. Dade-Robertson, *Furness Abbey: Romance, Scholarship and Culture* (Lancaster: Centre for North-West Regional Studies, 2000); and J. Wood, 'Visualizations of Furness Abbey: from romantic ruin to computer model', *English Heritage Historical Review* 3 (2008), pp. 8–35.

2 Key references to the history and archaeology of the site are T. West, *The Antiquities of Furness: Or An Account of the Royal Abbey of St Mary ...* (London: T. Spilsbury,

Despite its remoteness, the abbey did much to welcome and accommodate pilgrims and other visitors throughout the medieval period. The first travellers had an exciting, dramatic and sometimes treacherous time if it. This was because the obvious line of approach from the south was by crossing the hazardous tidal sands of Morecambe Bay; passing, in the allegorical words of Norman Nicholson, 'into an amphibian world that was neither earth nor water ... [where] ... the crossing gave [the visitor] an experience of forgetfulness, of oblivion, of detachment from the life of everyday ... like drowning and being revived'.[3] For some unfortunate travellers, however, drowning rather than revival was all too common. Indeed, there were so many victims of the perilous shifting quicksands that in 1377 Edward III granted the abbot of Furness the right to appoint a local coroner to investigate the number of deaths.[4] The first official Guide to the Sands was not appointed until 1536. Subsequent travellers had little option but to continue to use this Bay route, on foot, on horseback or availing themselves of a regular service of horse-drawn coaches and 'diligences', until the sands were by-passed by the turnpike, completed in 1820, and the Ulverston and Lancaster Railway, opened in 1857.[5]

Those caught out by the speed of tide on the Leven Sands, considered to be the most dangerous in the Bay, could take refuge, on the rocky outcrop known as Chapel Island, celebrated by William Wordsworth in his semi-autobiographical poem, *The Prelude*, published in 1805:

As I advanced, all that I saw or felt
Was gentleness and peace. Upon a small
And rocky island near, a fragment stood
(Itself like a sea rock) the low remains

1774) – see also a new edition with additions by W. Close (Ulverston: George Ashburner, 1805); T.A. Beck, *Annales Furnesienses, History and Antiquities of the Abbey of Furness* (London: Payne and Foss, 1844); J. Richardson, *Furness, Past and Present: Its History and Antiquities* (Barrow-in-Furness: J. Richardson, 1880 rev. edn, 1908), esp. pp. 117–41; W.H.St J. Hope, 'The Abbey of St Mary in Furness, Lancashire', *Trans of Cumberland and Westmorland Antiq. and Archaeol. Soc.*, old ser., 16 (1900), pp. 221–302 – reprinted as a separate book under the same title (Kendal: Titus Wilson, 1902); and J.C. Dickinson, 'Furness Abbey: an archaeological reconsideration', *Trans Cumberland and Westmorland Antiq. and Archaeol. Soc.*, new ser., 67 (1967), pp. 51–80. For a summary of current thinking see the latest guidebook: S. Harrison, J. Wood and R. Newman, *Furness Abbey, Cumbria* (London: English Heritage, 1998).

3   N. Nicholson, *The Lakers: The Adventures of the First Tourists* (London: Robert Hale, 1955), p. 67.

4   W. Farrer and J. Brownbill (eds), *The Victoria History of the County of Lancaster*, vol 2 (London: Archibald Constable, 1908), p. 119, fn 61.

5   Nicholson, *The Lakers*, p. 68. Today, cross-Bay walks are still a popular recreational activity guided by the Queen's Guide to the Sands. The present appointee is Cedric Robinson who became the 25th guide in 1963: C. Robinson, *Sandman* (Ilkley: Great Northern Books, 2009).

(With shells encrusted, dark with briny weeds)
Of a dilapidated structure, once
A Romish chapel, where the vested priest
Said matins at the hour that suited those
Who crossed the sands with ebb of morning tide.
Not far from that still ruin all the plain
Lay spotted with a variegated crowd
Of vehicles and travellers, horse and foot,
Wading beneath the conduct of their guide
In loose procession through the shallow stream
Of inland waters; the great sea meanwhile
Heaved at safe distance, far retired.[6]

The earliest known description of the abbey from the late eighteenth century is by the Revd William Gilpin who made a fleeting visit in 1772, observing that the abbey seemed

> to have been constructed in a good style of Gothic architecture; and has suffered, from the hand of time, only such depredations as picturesque beauty requires ... The proprietor of this noble scene is lord George Cavendish, who is a faithful guardian of it; preventing, as I am informed, any farther depredations.[7]

These words appeared in Gilpin's popular and influential *Observations Relative Chiefly to Picturesque Beauty* ... published in 1786, but it was the Lake District guidebooks of locally based writers like Fr Thomas West, and especially Wordsworth, which did much more to publicise the 'celebrated ruins'.[8]

---

6   W. Wordsworth, *The Prelude, or Growth of a Poet's Mind (Text of 1805)*, ed. E. de Selincourt, corrected by S. Gill, 2nd edn (Oxford: Oxford University Press, 1970), book tenth, lines 553–68. The event took place in 1794. The chapel on Chapel Island belonged to Conishead Priory not Furness Abbey as is sometimes stated.

7   W. Gilpin, *Observations Relative Chiefly to Picturesque Beauty, Made in the Year 1772, on Several Parts of England, Particularly the Mountains and Lakes of Cumberland and Westmoreland* (London: R. Blamire, 1786), pp. 158–9.

8   The earliest known ground plan of the abbey was made by West in about 1770, published in 1774: West, *The Antiquities of Furness*. T. West, *A Guide to the Lakes: dedicated to the lovers of landscape studies, and to all who have visited, or intend to visit the Lakes in Cumberland, Westmorland and Lancashire. By the author of the Antiquities of Furness* (London: Richardson & Urquhart and Kendal: W. Pennington, 1778) – West's *Guide* ran through at least twelve editions between 1778 and 1833; Wordsworth's *Guide* was first published anonymously in J. Wilkinson, *Select Views in Cumberland, Westmoreland, and Lancashire* (London: Ackermann, 1810) and went through a further four editions before reaching its final form in 1835: W. Wordsworth, *Guide through the District of the Lakes*, 5th edn (Kendal: Hudson & Nicholson and London: Longman & Co, 1835).

As a young man Wordsworth had a close affinity with the abbey, describing it 'as the boundary of his child-world, the object of a forbidden excursion'.[9] In *The Prelude*, he recalls an unruly school boy adventure (in about 1780), when he and his friends hire horses and race them along the sands near the abbey, and then into the abbey itself:

> … within the Vale
> Of Nightshade, to St. Mary's honour built,
> Stands yet a mouldering pile with fractured arch,
> Belfry, and images, and living trees;
> A holy scene! Along the smooth green turf
> Our horses grazed. To more than inland peace,
> Left by the west wind sweeping overhead
> From a tumultuous ocean, trees and towers
> In that sequestered valley may be seen,
> Both silent and both motionless alike;
> Such the deep shelter that is there, and such
> The safeguard for repose and quietness.
>
> Our steeds remounted and the summons given,
> With whip and spur we through the chauntry flew
> In uncouth race, and left the cross-legged knight,
> And the stone-abbot, and that single wren
> Which one day sang so sweetly in the nave
> Of the old church, that, though from recent showers
> The earth was comfortless, and, touched by faint
> Internal breezes, sobbings of the place
> And respirations, from the roofless walls
> The shuddering ivy dripped large drops, yet still
> So sweetly 'mid the gloom the invisible bird
> Sang to herself, that there I could have made
> My dwelling-place, and lived for ever there
> To hear such music.[10]

A slightly later visit, made by the popular Gothic novelist Mrs (Ann) Radcliffe in the summer of 1794, elegantly embodies the late eighteenth-century picturesque view and qualities of the abbey (Figure 11.1):

---

9   K. Hanley, 'The Stains of Time: Ruskin and Romantic Discourses of Tradition', in M. Wheeler and N. Whiteley (eds), *The Lamp of Memory: Ruskin, Tradition and Architecture* (Manchester: Manchester University Press, 1992), p. 102.

10   Wordsworth, *The Prelude*, book second, lines 103–28.

**Figure 11.1   A view of the abbey's north transept drawn by Thomas Hearne
in 1777, from a sketch made by Joseph Farrington, engraved in
1779 and published by Hearne and William Byrne in 1786**

Source: © T. Hearne and W. Byrne, *Antiquities of Great Britain, Illustrated in Views of
Monasteries, Castles and Churches now Existing* (London: T. Hearne and W. Byrne, 1786),
pl. XVI (The author's collection).

The deep retirement of its situation, the venerable grandeur of its gothic
arches and the luxuriant yet ancient trees, that shadow this forsaken spot,
are circumstances of picturesque and, if the expression may be allowed, of
sentimental beauty, which fill the mind with solemn yet delightful emotion. ...
the character of the deserted ruin is scrupulously preserved in the surrounding
area; no spade has dared to level the inequalities which fallen fragments have
occasioned in the ground, or shears to clip the wild fern and underwood that
overspread it; but every circumstance conspires to heighten the solitary grace
of the principal object, and to prolong the luxurious melancholy which the view
of it inspires.[11]

---

11   A. Radcliffe, *A Journey Made in the Summer of 1794 Through Holland and
the Western Frontier of Germany with a Return Down the Rhine, to which are added*

Three years after the visit of Mrs Radcliffe, came that by Johnson Grant, a young gentleman of the University of Oxford, in company with some college friends. Grant's party was on the last leg of a three-week tour of the north of England, ending in the Lakes in July 1797. Like many before them, their journey began by crossing the sands of Morecambe Bay:

> Early in the morning of the thirteenth, we left Lancaster, and went over the sands to Ulverston, in a coach which passes every day, when the tide is out. Every time the sea ebbs and flows, it alters the track of the coach, by carrying away or deposing sands. Thus guides are necessary, of which there are two at different places, paid by government, though not sufficiently to prevent them claiming perquisites from the passengers. Other carriages, and people on horseback, wait to follow the coach, and the whole proceed as in one caravan. ... Some weeks ago, three soldiers, crossing these sands, were overwhelmed by the water, two were drowned, and the third swam a mile and a quarter, quite exhausted, before he reached the shore.[12]

After staying overnight in Ulverston, the party set out for the abbey:

> We enter Furness abbey through a Gothic arched gateway, with a thick drapery of ivy, hanging gracefully down on one side. The afternoon was favourable for viewing it; as the sun was overspread with clouds, and a solemn calm reigned throughout the air, unbroken, but by the cawing of the rook, the drowsy hum of the flies, or the complaining of a haunted stream. ... The approach is lined by venerable old trees, which envelop the ruin in their awful listening gloom. The way is strewed with fragments of desolation. Reached, through these, the silent contemplative remains of the tall pile; and a train of ideas of the most serious nature rush upon the mind; melancholy from remembrance, calm with stillness, breathing 'love of peace and lonely musing'. We are struck with an extensive ruin, exactly in a proper state of decay to shew the depredations of time, without effacing the grandeur of which it once was.[13]

This last sentence is clearly inspired by Gilpin's *Observations* ... published a decade or so earlier.[14] But as in Wordsworth's lines and the passage from Mrs Radcliffe, Grant's splendid prose is almost equally concentrated on the abbey's flora and fauna, as to its decaying architecture. The 'irregularly crumbled' ruins

---

Observations during a Tour of the Lakes of Lancashire, Westmorland and Cumberland (London: G.G. and J. Robinson, 1795), pp. 487–8.

12   J. Grant, *Journal of a Three Weeks Tour in 1797, through Derbyshire, to the Lakes*, in W. Mavor, *The British Tourists, or Traveller's Pocket Companion, through England, Wales, Scotland, and Ireland*, vol. 4, 3rd edn, (London: Richard Phillips, 1809), p. 259.

13   Ibid., p. 263.

14   Gilpin, *Observations*, p. 158.

are 'begirt with ivy, and bestrewed with wild shrubs and high waving grass'. The noisy rook, flapping through the 'high and deep-wrought Gothic windows', affords 'a melancholy proof of desertion and desolation'. The impression is one of nature reclaiming this hallowed place. Indeed Grant describes the panorama as a 'hallowed grove'. But it is also a 'haunted' scene, charged with imaginative energy that summons remembrance and meaning:

> How grand must have been the deep-toned organ's swell, the loud anthem of a hundred voices, rolling through these roofs … Even now, when all these heaven-inspiring sounds have ceased, even now does memory recur to them; and fancy peoples the gloom with all its former inhabitants.[15]

A further picturesque description of the abbey, some twenty years later, is provided by the Revd Francis Edward Witts, who toured the Lakes for about a fortnight in August and September 1817, staying near Ulverston for some of this time. He visited the abbey on 1 September and urged others to do likewise:

> The ruins of Furness Abbey … well repay a visit, and the traveller may pass an hour with pleasure in rambling within the venerable inclosure, tracing new groupes of picturesque beauty, as the aged remains cluster among the foliage around.[16]

Witts appreciated the dilapidated ruins for their aesthetic appeal. For him, the ivy-clad 'time worn walls', afforded 'a fine subject for the pencil of the artist skilled in the portraiture of architectural antiquities'.[17] It also presented a rural idyll (Figure 11.2):

> the brook precipitates itself thro' the dell, in many places concealed beneath the foundations of the buildings, where it has been overarched. In one spot, where the vault had fallen in, leaving a large chasm over the water, we observed a fisherman throw his line into the hole with repeated success for trouts.[18]

---

15   Grant, *Journal of a Three Weeks Tour*, p. 264.

16   A. Sutton (ed.), *The Complete Diary of a Cotswold Parson: The Diaries of the Revd Francis Edward Witts, 1783–1854*, in 10 volumes, *Volume 2: The Curate and Rector: Diaries covering the period 1806–1825* (Stroud: Amberley Publishing, 2008), p. 228. I am grateful to Janet Martin for drawing this work to my attention.

17   Ibid. Edward Baines witnessed just such an artistic enterprise: 'As I was leaving the Abbey, a party of two ladies and a gentleman arrived for the purpose of sketching': E. Baines, *A Companion to the Lakes of Cumberland, Westmorland and Lancashire in a descriptive account of a Family Tour and excursion on horseback and on foot* (London, Hurst, Chance & Co, 1829), p. 245.

18   Sutton, *The Complete Diary of a Cotswold Parson*, p. 227.

The intrusion into this prospect of the manorial residence of Lord George Cavendish (the abbey's owner) was disliked by Witts as having 'a modern and vulgar look amidst the fine Gothic ruins'.[19] Witts also refers to an adjacent cluster of tenanted farm buildings. These were evidently still being used in about 1840, when Thomas Alcock Beck was researching his scholarly book, *Annales Furnesienses* ... . Beck complained that the 'tone of gentle enchanting melancholy ... is harshly interrupted by the various discordancies of agricultural employment'.[20]

Figure 11.2   A view of the abbey from the south by the architect and prolific illustrator Thomas Allom, engraved by E. Challis and published in 1833

Source: © Fisher, Son and Co. of London (The author's collection).

But it was an intrusion of a wholly different nature and scale that was to place Furness Abbey firmly on the map – more specifically, on the railway map – namely, the construction of the Furness Railway and the decision to build a station and hotel right next to the abbey. In fact it could have been even more insensitive as the original

---

19   Ibid. The post-Dissolution manor house probably incorporated some upstanding medieval work. The Cavendishes lived at several miles away at Holker Hall and only visited occasionally.

20   Beck, *Annales Furnesienses* ... p. 367.

survey, by the eminent engineer James Walker, had the railway line passing through part of the abbey itself. In the end the line was diverted through a tunnel to skirt the ruins.[21] Since the death of Lord George Cavendish in 1834, ownership of the abbey had passed to his grandson, William Cavendish, second earl of Burlington (later seventh duke of Devonshire). William owned mineral royalties, which enabled him to profit from the lucrative local deposits of haematite iron ore, and unsurprisingly promoted and fashioned the railway to suit his own purposes. He also stood to gain from the sale of the land for the track and the manor house buildings for the station and hotel. Predictably, from 1848, he became chairman of the Furness Railway Company.[22]

The arrival of the railway did not win universal approval. Wordsworth's wife Mary, in a letter to Henry Crabb Robinson dated 21 June 1845, complained that on a recent excursion three acquaintances had 'had the pain (tho it was a picturesque appearance) of seeing the Old Abbey occupied by the 'Navys' at their meal, who are carrying a rail-way, so near to the East window that from it Persons might shake hands with the Passengers!!'[23] In a sonnet composed on the same day as his wife's letter, Wordsworth himself reveals his strong disapproval, contrasting the reverence of the railway workers with the landowners who allowed the abbey grounds to be invaded:

Well have yon Railway Labourers to THIS ground
Withdrawn for noontide rest. They sit, they walk
Among the Ruins, but no idle talk
Is heard; to grave demeanour all are bound;
And from one voice a Hymn with tuneful sound
Hallows once more the long-deserted Quire
And thrills the old sepulchral earth, around.
Others look up, and with fixed eyes admire
That wide-spanned arch, wondering how it was raised,
To keep, so high in air, its strength and grace:
All seem to feel the spirit of the place,
And by the general reverence God is praised:
Profane Despoilers, stand ye not reproved,
While thus these simple-hearted men are moved?[24]

---

21   *Plan and Section of an Intended Railway from Rampside and Barrow to Dalton and Ulverstone and to the Existing Railway at or near Sandside in the Parish of Kirkby Ireleth in the County Palatine of Lancaster*, surveyed by J.R. Wright under the direction of Walker and Burgess Engineers, 1843 (London, House of Lords Record Office, 1844), F2, Furness Railway.

22   J. Simmons, *The Victorian Railway* (London: Thames and Hudson, 1991), pp. 162–3; M. Andrews, *The Furness Railway in and around Barrow* (Pinner: Cumbrian Railways Association, 2003), pp. 9, 13. On the economic exploitation of the Cavendish estates and the development of Barrow, see J.D. Marshall, *Furness and the Industrial Revolution* (Barrow-in-Furness: Barrow-in-Furness Library and Museum Committee, 1958).

23   A.G. Hill (ed.), *The Letters of William and Dorothy Wordsworth, 7: The Later Years, 4, 1840–1853*, 2nd edn (Oxford: Oxford University Press, 1988), p. 679.

24   W. Wordsworth, *At Furness Abbey*, 21 June 1845.

Wordsworth's contempt is further evident in a letter to his nephew, Christopher Wordsworth Jr, written in August 1845: 'What do you think of a Railway being driven as it now is, close to the magnificent memorial of the piety of our ancestors? Many of the trees which embowered the ruin have been felled to make way for this pestilential nuisance.'[25] These words were later echoed by the author James Payn:

> A railway was made to violate the slumberous repose of the Valley of Nightshade, and that with every circumstance of Gothic ferocity to enhance the crime; its sleepers were laid down within a few feet of the spot where the mailed Barons of Kendal had hoped to find an undisturbed resting-place.[26]

The Furness Railway line and the Furness Abbey Station opened in 1847. Whilst many at the time clearly considered the construction of the railway a notoriously insensitive decision, it did result in the abbey almost certainly becoming the first ancient monument in Britain to have its own railway station.[27] Although the railway was initially principally a mineral line, it did provide a small-scale passenger service at this early date, with connections to the south via cross-Bay steamers from Fleetwood and Lancaster. The railway company had purchased the steamer *Helvellyn* in 1847 to operate out of Fleetwood,[28] while the steamers *Duchess of Lancaster* and *John Denistown* operated out of Lancaster. An anonymous writer recounts a trip to the abbey by the steamer *Duchess of Lancaster* in July 1848, in the form of a romantic poem which contrasts the decay of the ruins with the newness of the station: 'The Abbey on the one hand see /All the mouldering to decay; /And on the other, glittering bright, /A railway station, gay.'[29]

It would appear, however, that some of the railway works were still being completed in 1849, when the abbey was visited by Queen Victoria. The Queen and her entourage arrived by carriage rather than by train, and observed and even overheard the railway workers. The scene was recorded by Lady Augusta Elizabeth Frederica Bruce (later Stanley), at the time lady-in-waiting to the Queen's mother, the Duchess of Kent, and later, from 1861, one of the Queen's most loyal and beloved ladies-of-the-bedchamber. Lady Augusta was the fifth

---

25   Hill, *The Letters of William and Dorothy Wordsworth*, p. 700.

26   Quoted in F. Ross, *The Ruined Abbeys of Britain* (London: William Mackenzie, c.1882), p. 213.

27   The only other station specifically built to serve an abbey as a tourist attraction appears to be Neath Abbey Station on the Swansea & Neath Railway, opened in 1863: G. Biddle, *Victorian Stations: Railway Stations in England & Wales 1830–1923* (Newton Abbot: David & Charles, 1973), p. 207. Tintern Station, on the Wye Valley Railway, located one mile north of Tintern Abbey, was very much the abbey station, with excursions run for those wanting to visit the ruins. Again this is much later, opening in 1876.

28   K.J. Norman, *The Furness Railway, 2: Locomotives, Ships, Excursions and Miscellanea* (Kettering: Silver Link Publishing, 2001), p. 27.

29   *A Pleasure Trip to Furness Abbey by the Steamer, 'The Duchess of Lancaster', 13 July 1848* (Lancaster, 1895), quoted in Dade-Robertson, *Furness Abbey*, p. 37.

daughter of Thomas Bruce, seventh earl of Elgin and Kincardine, who had brought the Elgin Marbles back from the Parthenon in Athens. Like her father, she was intelligent and cultured, with an entertaining personality, leaving 'two volumes of letters, brimming with wit, sharpness of observation, and a healthy sense of humour'.[30] Her amusing letter recounting the royal visit to 'Barrow Abbey' reads as follows:

> We started in an open carriage with the dear Queen and had a charming drive to the Abbey which far surpassed anything I had anticipated. So beautiful it is and so extensive – I never saw anything of the kind so gigantic in its proportions. Ferns and ivy growing in the chinks of the walls make it look lovely even in its desolation, and there are, scattered about in every direction, perfect gems of architecture. A rapid brook runs through the grounds, and must in former days have blended its voice with the swelling sounds of the organ, and joined in the hymn of praise. The sun shone brightly and brought out every beauty, but the noise of the workmen and the intruding railway disturbed the peacefulness of the scene and jarred painfully on our feelings. ... One of the workmen was overheard by Mrs Gwyllym to say to his companion, in the broad Lancastrian dialect, 'Well, if this here belonged to me, why I'd try if I could to build it up again as it was before'. Was not that a nice sentiment from a railway labourer?[31]

It would be almost another decade before the line's expansion, to link up with Ulverston and Lancaster Railway in 1857, brought to an end the relative isolation of this part of Lancashire and the vagaries of its transport links (Figure 11.3).[32] With the increased rail traffic now possible, the abbey quickly became a popular tourist destination, a situation foretold by the Lancaster-based architect and medievalist Edmund Sharpe:

> whilst we cannot help lamenting the inroad which the [railway] innovation has made upon the former appearance of this interesting valley, we must not forget that we owe our visit here to-day to the facilities which it offers, and which have been rendered available to thousands, to whom this interesting spot would otherwise have been inaccessible.[33]

---

30   H. Rappaport, *Queen Victoria: A Biographical Companion* (Oxford: ABC-Clio, 2003), p. 81.

31   The Dean of Windsor (A.V. Baillie) and H. Bolitho (eds), *Letters of Lady Augusta Stanley: A Young Lady at Court, 1849–1863* (London: Gerald Howe, 1927), pp. 24–5.

32   *Illustrated London News* (5 September 1857), pp. 245–6: 'The line would undoubtedly be most serviceable to the district, and this would be remembered as a great day for "Lonsdale north of the sands"'.

33   E. Sharpe, 'The ruins of the Cistercian monastery of St Mary in Furness', *Journal of the British Archaeological Association* 6 (1851), pp. 315–6.

DEJEUNER IN FURNESS ABBEY, IN CELEBRATION OF THE OPENING OF THE ULVERSTONE AND LANCASTER RAILWAY.

Figure 11.3   An illustration of the grand dinner for 300 guests, held in a giant marquee at the abbey, to celebrate the opening of the Ulverston and Lancaster Railway in 1857

Source: © *London Illustrated News* (The author's collection).

Sharpe's approval was to be expected, as his architectural practice, with Edward G Paley, had been responsible, on behalf of the Furness Railway Company, for the conversion of the manor house to the Furness Abbey Hotel in 1847. As a pioneer 'country-house hotel' it is considered to be the first railway hotel in Britain aimed solely at the tourist market.[34] In the late 1860s, Paley carried out additions and alterations to the hotel and adjacent station, connecting the two with a covered walkway. Paley also built Abbotswood House in 1857 on the hill overlooking the station. This was the home of Sir James Ramsden, General Manager and later Managing Director of the Furness Railway Company. The station was provided with a special siding for Ramsden's own saloon, in which he travelled daily, to and from Barrow, so as to avoid other people.[35]

---

34   Simmons, *The Victorian Railway*, p. 40. C. Wolmar, *Fire & Steam: How the Railways Transformed Britain* (London: Atlantic Books, 2007), p. 120. On this theme, see J. Simmons, 'Railways, hotels and tourism in Great Britain, 1839–1914', *Journal of Contemporary History* 19, 2 (1984), pp. 201–22. The North Euston, the resort hotel in Fleetwood built in 1841, must have a rival claim, though admittedly it was built as a stop-over on the way to Scotland before the railway was continued over Shap, rather than as a destination hotel *per se*.

35   J. Price, *Sharpe, Paley and Austin: A Lancaster Architectural Practice: 1836–1942* (Lancaster: Centre for North-West Regional Studies, 1998), pp. 70, 72; D. Joy, *A Regional*

As Sharpe predicted, the railway did much to boost the abbey's tourist trade throughout the second half of the nineteenth century, making an attractive day out possible for people of the northern industrial towns, and visitors from further afield. The site even welcomed its first recorded American visitor as early as July 1855, although he arrived by horse-drawn carriage. This was Nathaniel Hawthorne, the novelist, writer and United States consul, who came to England in 1853 for a consular term of four years. Hawthorne's visit clearly left him impressed by the abbey, as well as charmed by the railway and hotel:

> There is a railway station close by the ruins; and a new hotel stands within the precincts of the abbey grounds; and continually there is the shriek, the whiz, the rumble, the bell-ringing, denoting the arrival of the trains; and passengers alight, and step at once (as their choice may be) into the refreshment-room, to get a glass of ale or a cigar, – or upon the gravelled paths of the lawn, leading to the old broken walls and arches of the abbey ... arches loftier than I ever conceived to have been made by man ... arches, through which a giant might have stepped, and not needed to bow his head, unless in reverence to the sanctity of the place. ... I liked the effect of so many idle and cheerful people, strolling into the haunts of the dead monks, and going babbling about, and peering into the dark nooks; and listening to catch some idea of what the building was from a clerical-looking personage, who was explaining it to a party of his friends. I don't know how well acquainted this gentleman might be with the subject; but he seemed anxious not to impart his knowledge too extensively, and gave a pretty direct rebuff to an honest man who ventured an inquiry of him.[36]

As Hawthorne readily admits, it was still virtually impossible to describe and understand the ruins in any meaningful way: 'I have made a miserable botch of this description; it is no description, but merely an attempt to preserve something of the impression it made on me, and in this I do not seem to have succeeded at all'.[37] Hawthorne's defeatism at losing his descriptive prowess was perhaps unsurprising, as he divulges that he had never troubled 'to form any distinct idea of what an abbey or monastery was'.[38]

As United States consul, Hawthorne was based in Liverpool. It was here, in 1869, that another notable American tourist, a young Theodore 'Teddy' Roosevelt,

---

History of the Railways of Great Britain, Vol. 14: The Lake Counties, 2nd edn (Newton Abbot: David and Charles, 1990), pp. 114, 124. Abbotswood was demolished in the 1960s.

36   N. Hawthorne, *Our Old Home, and English Notebooks*, in *The Complete Works of Nathaniel Hawthorne*, with introductory notes by G.P. Lathrop, 8 vols (London: Kegan Paul, Trench and Company, 1883), vol. 2, pp. 10–14. See also J. O'Donald Mays, *Mr Hawthorne Goes to England: The Adventure of a Reluctant Consul* (Burley: New Forest Leaves, 1983).

37   Hawthorne, *Our Old Home*, p. 14.

38   Ibid., p. 13.

the future 26th President, disembarked at the start of his family's European Grand Tour. 'Teedie', as he was nicknamed, was only aged ten at the time, and a sickly and asthmatic child with frequent ailments. Indeed, one of the reasons for the tour was to improve his health. Despite his illnesses, however, Teedie was hyperactive and often mischievous with an insatiable hunger for adventure. He displayed these characteristics on the family's visit to Furness Abbey.

The visit is described in detail by David McCullough in his book *Mornings on Horseback*. The abbey, whose ruins were 'considered among the choicest of all' was to be the family's 'maiden encounter with Europe's truly ancient past ... on their own as tourists for the first time'. The party comprised Teedie's father, Theodore Roosevelt Sr, his mother, Martha or 'Mittie', his older sister Anna, Teedie himself, his younger siblings Elliot and Corinne, and the children's nursemaid. On arrival in England, the party increased to eight with the addition of a newly hired valet.[39] After leaving Liverpool by train on 2 June, they alighted at dusk at Furness Abbey Station, and spent a night at the Furness Abbey Hotel. It was only in the shining light of the next morning that 'the scale and power of the place burst upon them':

> What had been the floor of the abbey was now vivid green turf speckled with bluebells and buttercups and there were as yet no restrictions as to where one could walk or climb; everything was open to all comers and for children, a glorious playground. ... The three small children ... raced ahead, through archways, up broken stairs, then up a spiral stairway to the belfry ... From the cracked stone helmet of [an] effigy Mittie picked a dandelion. Thrilled by the whole romantic spell of the place she found herself 'gazing at the wide open windows and thinking how the glorious light must have streamed through ... over the high altar [and] down upon the monks as they sang their solemn chants or the moonbeams when at their midnight devotions'.[40]

It is not recorded if Hawthorne or the Roosevelt family purchased an abbey guidebook, or were escorted around the ruins by the abbey's official uniformed guide (probably not in Hawthorne's case, given his confession). Guidebooks and an official guide, however, were available at this time. Indeed, the first in a succession of guidebooks was produced as early as 1845. Initially, these were written anonymously,[41] but later

---

39   D. McCullough, *Mornings on Horseback: The Story of an Extraordinary Family, a Vanished Way of Life and the Unique Child Who Became Theodore Roosevelt*, 2nd edn (New York: Simon & Schuster, 2001), pp. 78–80.

40   Ibid., 79–80.

41   *A Handbook to the Abbey of St Mary of Furness, in Lancashire* (Ulverston: S. Soulby, 1845); *A Guide to the Ruins of Furness Abbey ...* (Ulverston: D. Atkinson, 1846); *A Handbook to Furness Abbey ...* (Ulverston: J. Jackson, 1847) – rev. *c.*1860; *A Guide from Blackpool and Fleetwood to Furness Abbey ...* (Blackpool: W. Porter, 1847); *A Guide Through the Ruins of Furness Abbey ...* (Ulverston: D. Atkinson, 1854). See also F. Evans, *Furness and Furness Abbey; Or a Companion through the Lancashire Part of the Lake*

editions reveal their authors to have been Thomas Alcock Beck and especially J. P. Morris, whose *Handy Guide* continued to be revised into the 1920s.[42] It would appear that Furness Abbey was one of the first ancient monuments in Britain to employ an official guide. The post was initially occupied by a John Fisher from about 1850 until 1873. There is a splendid photograph of him in his uniform, sitting in the vestibule of the chapter house (Figure 11.4).[43] Fisher was succeeded by William Poole from 1873 until 1891, while the third guide, Jesse Turner, seems to have been in post until about 1934. All three guides appear to have been employed by the Furness Railway Company, rather than the Cavendishes, the state presumably taking over Turner's employment from 1923. The guides were also provided with domestic accommodation in a converted late medieval building, now known as Abbey Park Cottage, and with a heated half-timbered shelter at the entrance to the site.[44]

Very little is known about John Fisher or William Poole, or of the role they performed. Poole's obituary notice records that he 'was full of information' and had 'picked up many interesting facts which were acceptable to antiquarians and archaeologists'.[45] More details are available for Jesse Turner. He is known to have assisted Sir William H. St John Hope with the first systematic excavations and archaeological assessment of the site between 1896 and 1898, and again between 1990 and 1901. As a result, Turner became very knowledgeable about the abbey and was responsible for at least one revision of Morris' *Handy Guide*.[46] There is a wonderful postcard of him, produced by Atkinson's of Ulverston, resplendent in

---

District (Ulverston: D. Atkinson, 1842), pp. 195–214; C. M. Jopling, *Sketch of Furness and Cartmel, Comprising the Hundred of Lonsdale North of the Sands* (London: Whittaker & Co., 1843), pp. 105–44; J. Payn, *Furness Abbey and its Neighbourhood* (Windermere: J. Garnett, *c.*1858); H. Barber, *A Tourist's Guide to Furness Abbey and its Vicinity* (Ulverston: J. Atkinson, *c.*1865); H. Barber, *Shaw's Tourists' Picturesque Guide to Furness Abbey ...*, 8th edn (London: Norton and Shaw, *c.*1878). Dade-Robertson, *Furness Abbey*, chapter 4 examines in detail local guidebooks to the abbey and the surrounding area.

42   J.P. Morris, *A Handy Guide to the Ruins of Furness Abbey* (Ulverston: J. Atkinson, 1875).

43   Glass lantern slide by A. Pettitt entitled 'Furness Abbey Guide *c.*1850', Barrow Record Office and Local Studies Library (S.B. Gaythorpe collection).

44   William Poole joined the staff of the Furness Railway Company in January 1851 and was given rent free accommodation as the abbey guide from November 1873: The National Archives, RAIL 214/97/47 POOL. I am grateful to Stephen Fullard for what information exists on the abbey guides, and also to Janet Martin for drawing my attention to the existence of a part-time guide, Myles King, who occupied Abbey Park Cottage in 1851: Lancashire Record Office, DDCa/1/155; Census 1851. For Abbey Park Cottage see J. Wood, 'Furness Abbey: A Case Study in Monastic Secularization', in T. N. Kinder (ed.), *Perspectives for an Architecture of Solitude. Essays on Cistercians, Art and Architecture in Honour of Peter Fergusson* (Turnhout: Brepols, Medieval Church Studies 11 & Cîteaux: Commentarii Cistercienses, Studia et Documenta 13, 2004), pp. 377–85.

45   *The Barrow News*, 20 October 1891.

46   Morris, *Handy Guide*, rev. edn by J. Turner (post-1923).

Figure 11.4 A photograph of John Fisher, the abbey's first official guide
Source: © Cumbria Archive and Local Studies Centre, Barrow, courtesy of Stephen Fullard.

his official uniform and standing in the infirmary hall.[47] Another photograph, dated 1905, finds him standing next to four schoolgirls seated on the sedilia.[48]

In the early 1860s many of the tomb effigies and other vulnerable decorative sculptures were relocated from the church to the infirmary chapel. The chapel also acquired an iron-railing gate and in effect became the site museum, patrolled by the abbey guides. The first souvenirs also appear at this time, some perhaps sold by the abbey guides. 'Carte de visite' albumen photograph cards were produced by Alfred Pettitt of Keswick and Gilsland among others, later replaced by a proliferation of postcards of the abbey and hotel, issued by various companies including Raphael Tuck and Son for the Furness Railway Company, Valentine and

47   J.D. Marshall and M. Davies-Shiel, *Victorian and Edwardian Lake District, from Old Photographs* (London: B.T. Batsford, 1976), no. 86.

48   J. Garbutt and J. Marsh, *Around Barrow-in-Furness in Old Photographs* (Stroud: Sutton Publishing, 1993), p. 129.

Sons Ltd, McCorquodale and Co. and the local photographer Raymond Sankey. Other souvenirs included various ceramics, match holders, spoons, pin trays (Figure 11.5), brooches (Figure 11.6) and even a silver-plated lobster pick.

Figure 11.5   A pewter pin tray souvenir from Furness Abbey
Source: © The author.

Figure 11.6   A brooch, carved from bog oak, depicting the east end of the church at Furness Abbey
Source: © The author.

The late 1870s and early 1880s saw the first attempts at de-vegetation, cleaning and restoration of the abbey ruins. John Ruskin was not an enthusiastic supporter of this work: 'I had almost rather see Furness or Fountains Abbey strewed in grass-grown heaps by their brook-sides, than in the first glow and close setting of their fresh-hewn sandstone'.[49] Nor was Ruskin endeared to the clientele of the Furness Abbey Hotel, with its freely available refreshments. This, from a letter dated October 1871:

> I was waiting last Saturday afternoon on the platform of the railway station at Furness Abbey (the station itself is tastefully placed so that you can see it, and nothing else but it, through the east window of the Abbot's Chapel, over the ruined altar), and a party of the workmen employed on another line ... were taking Sabbatical refreshment at the tavern recently established at the south side of the said Abbot's Chapel. Presently, the train whistling for them, they came out in a highly refreshed state, and made for it as fast as they could by the tunnel under the line, taking very long steps to keep their balance in the direction of motion, and securing themselves, laterally, by hustling the wall, or any chance passengers. They were dressed universally in brown rags, which, perhaps, they felt to be the comfortablest kind of dress; they had, most of them, pipes, which I really believe to be more enjoyable than cigars; they got themselves adjusted in their carriages by the aid of snatches of vocal music, and looked at us ... with supreme indifference, as indeed at creatures of another race; pitiable, perhaps – certainly disagreeable and objectionable – but, on the whole, despicable, and not to be minded. We, on our part, had the insolence to pity them for being dressed in rags, and for being packed so close in the third-class carriages: ... and when a thin boy of fourteen or fifteen, the most drunk of the company, was sent back staggering to the tavern for a forgotten pickaxe, we would, any of us, I am sure, have gone and fetched it for him, if he had asked us. For we were all in a very virtuous and charitable temper: we had had an excellent dinner at the new inn, and had earned that portion of our daily bread by admiring the Abbey all the morning. So we pitied the poor workmen doubly – first, for being so wicked as to get drunk at four in the afternoon; and secondly, for being employed in work so disgraceful as throwing up clods of earth into an embankment, instead of spending the day, like us, in admiring the Abbey.[50]

Three years later, in 1874, Ruskin's opposition to the proximity of the railway to Furness Abbey became the subject of public notoriety when he declined the

---

49   J. Ruskin, Preface to *Notes by Mr Ruskin on Samuel Prout and William Hunt, in Illustration of a Loan Collection of Drawings Exhibited at the Fine Arts Society's Galleries, in 1879–80* (London: The Fine Art Society, 1880).

50   J. Ruskin, *Fors Clavigera*, Letter 11, 'The Abbot's Chapel', 15 October 1871; quoted by J.S. Dearden, 'Ruskin's thoughts on Furness Abbey', in *Facets of Ruskin: Some Sesquicentennial Studies* (London: Charles Skilton, 1970), p. 78.

award of a medal by the Royal Institute of British Architects in protest against the despoiling of beautiful buildings by architects. Ruskin quoted one example in Britain and three in Italy. The construction of the railway at Furness Abbey was the British one.[51] This act of railway vandalism was later immortalised in a cartoon published in *Punch*.[52] Ruskin's hostility continued to be aroused in succeeding years. In 1875, he prophesised sardonically a 'glorious England of the future; in which there will be no abbeys (all having been shaken down, as my own sweet Furness is fast being, by the luggage trains)'.[53] In 1876, he castigated 'the navvies of Furness' as being 'a fallen race, fit for nothing but to have dividends got out of them, and then be damned'.[54] In his Oxford Lectures of 1883, Ruskin castigated their employers even more, taking issue with 'railroad … companies, that … propose to render beautiful places more accessible'. He again cites Furness Abbey as one of four of 'the most beautiful and picturesque subjects once existing in Europe', now defaced by the proximity of railways. 'Since these improvements have taken place', he laments, 'no picture of any of these scenes has appeared by any artist of eminence, nor can any in future appear. Their portraiture by men of sense or feeling has become for ever impossible.'[55]

Despite Ruskin's disparagements, his 'own sweet Furness', his place of private pleasure and contemplation, was now, because of the railway, 'rendered available to thousands', to borrow Sharpe's prophetic words. Ruskin died in 1900, and so escaped witnessing the further expansion of the Furness Abbey Hotel, and enlargement of the station, which the turn of the century brought (Figure 11.7). The 36-bedroom hotel, the flagship of the Furness Railway Company, now offered conference accommodation. The United Association of Bakers and Confectioners held their conference there in about 1910 (Figure 11.8),[56] and the hotel also hosted a Furness Railway Ambulance competition in about 1906.[57] This was the era when the railway company, under the entrepreneurial Alfred Aslett, was starting to benefit from its astute policy of developing the tourist appeal of the area, including cross-bay steamers linking up with Lake District tours, to counterbalance the decline – already – in iron ore revenues.

---

51   Dearden, 'Ruskin's thoughts on Furness Abbey', pp. 78–9.

52   5 February 1876: see J. Abse, *John Ruskin: The Passionate Moralist* (New York: Alfred A. Knopf, 1981), pp. 156–7.

53   Ruskin, *Fors*, Letter 56, August 1875. The vibration caused by the heavy mineral trains was probably more damaging.

54   Ibid., Letter 64, April 1876.

55   J. Ruskin, *The Art of England: Lectures Given in Oxford* (Orpington: George Allen, 1884), p. 265. For Ruskin on railways see K. Hanley and J.K. Walton, *Constructing Cultural Tourism: John Ruskin and the Tourist Gaze* (Bristol: Channel View Publications, 2010), chapter 5; also J.M. Richards, 'The role of the railways', in M. Wheeler (ed.) *Ruskin and Environment* (Manchester: Manchester University Press, 1995), pp. 123–43.

56   The National Archives, RAIL 214/91/14.

57   Barrow Record Office and Local Studies Library (S.B. Gaythorpe collection), Z 2915.

Figure 11.7   The Furness Abbey Hotel and station.
Note: The hotel was built in red sandstone to harmonise with the abbey buildings, while the half-timbered station was capped with a steepled clock tower and finials
Source: © English Heritage.

Figure 11.8   The United Association of Bakers and Confectioners conference, photographed at Furness Abbey *c*.1910
Source: © The National Archives.

Visitors and conference delegates were perhaps enticed by advertisements spuriously claiming that the hotel was 'In the Centre of Lakeland'. It could not be less central. One such advertisement, of 1907, also claims that the hotel is 'the favourite resort of the artist, antiquary, and lovers of the picturesque', offering an 'enlarged entrance hall with inglenook … perfect sanitation … electric light … handsome ball room … well-appointed billiard room … improved second class refreshment room … [and] … electric bell communication to the abbey guide'.[58] The hotel is prominently displayed in another Furness Railway advertisement of June 1916, this time promoted as 'The Centre for Lakeland'. Notice the subtle difference.[59] The place mythology is discontinued in a less tempting, but more geographically accurate, advertisement on the back cover of one of Turner's revisions of Morris' *Handy Guide*. The hotel is now 'a convenient break of journey for Westmorland and the Lakes' – a circuitous journey but we will let that pass – and 'situated on the outskirts of Barrow-in-Furness, which can be reached in a few minutes', which, as today's marketing managers would say, is perhaps rather too much information. The advertisement is undated but as the hotel is described as 'One of the LMS Hotels', it must post date the absorption of the Furness Railway Company into the London, Midland and Scottish Railway in 1923.[60]

1923 was also the year that Lord Richard Cavendish, younger brother of the ninth duke of Devonshire, placed the abbey in the guardianship of the state.[61] The site now became the responsibility of the Office of Works, under Sir Charles Peers, Chief Inspector of Ancient Monuments, who immediately set in motion a robust programme of fabric intervention, stabilisation, consolidation and restoration which continued unabated until the early 1930s. The work was accompanied by a general 'tidying up' of the abbey grounds, consistent with the Office of Works' 'institutionised'

---

58   *Furness Abbey Hotel*, 1907, reproduced in Dade-Robertson, *Furness Abbey*, p. 82. See also an advertisement in *The Railway News*, 5 July 1913 and the *Tariff of the Furness Abbey Hotel, Barrow-in-Furness* (Empire Hotels, *c.*1909). Until 1913, the hotel was managed for the Furness Railway by the London catering firm of Spiers & Pond which was also responsible for the refreshment room at the station.

59   *Furness Railway: The Gateway to Lakeland*, 1916, reproduced in Dade-Robertson, *Furness Abbey*, p. 86. On this theme see Dade-Robertson, *Furness Abbey*, chapter 4 and Norman, *The Furness Railway*, 2, chapter 3.

60   *Furness Abbey Hotel* (post-1923), reproduced in Dade-Robertson, *Furness Abbey*, p. 89. Exaggerated claims persist. While writing this chapter, I received alluring publicity for an overnight stay for two in the Lake District from Groupon. Closer examination of the small print revealed that the 'good snooze' was to be had at the Abbey House Hotel, 'within rolling distance of Barrow-in-Furness, rubbing shoulders with the ruins of Furness Abbey', and in reality 'teetering at the foot of the Lake District … around a 30 minute trundle from Cumbria's peaks and puddles'. The Abbey House Hotel opened in 1986. Built in 1914 to the designs of Sir Edwin Lutyens, it was originally the home and guest house of the Chairman of Vickers Ship Building: http://www.abbeyhousehotel.com/.

61   Cavendish remained the owner, retaining the honorary title of 'Lay abbot of Furness'. The present owner is Hugh Cavendish, Richard's grandson.

regime of treatment for ruined sites, introduced by Peers.[62] Given this sustained level of public investment, it is perhaps surprising that it took until 1943 before the first HMSO official guidebook was produced; and that it was so inadequate.[63] It would be over twenty years before this deficiency was put right.[64] In the meantime, the Furness Abbey Hotel had closed in 1938, and was subsequently requisitioned by the military. Both the hotel and the station suffered bomb damage in 1941, and at the end of the war the hotel stood forlorn and neglected. Passenger services were withdrawn from the station in 1950 and the station buildings demolished two years later. Except for part of its north wing, the hotel was demolished in 1953.[65]

Through the picturesque and romantic movements of the late eighteenth and early nineteenth centuries, Furness Abbey was a remote but rewarding destination for an elite group of travellers and writers. From this period, and through their published works, it became assimilated into the Lake District and its tourism history. The coming of the railway brought the Furness peninsula closer, in travelling time, towards the orbit of the Lakes and stimulated further the growth in popular tourism at the abbey. For William Wordsworth and John Ruskin, the abbey was a site invested with strong personal associations, and they reserved some of their more scathing, forceful and (in the case of Ruskin) histrionic words for condemning the decision to build the railway within touching distance of the ruins. But as a railway destination the abbey, surprisingly, became something of a trend-setter in Britain, as almost certainly the first ancient monument to have its own railway station, the first railway hotel aimed solely at tourists and one of the first sites of its kind to employ an official guide.

---

62   J. Swarbrick, 'The Reparation of Furness Abbey', *National Ancient Monuments Review* 2, 5 (1929), pp. 204–8; 6 (1929), pp. 263–70. On this theme see A. Keay, 'The Presentation of Guardianship Sites', *Transactions of the Ancient Monuments Society* 48 (2004), pp. 7–20.

63   S.J. Garton, *Furness Abbey, Lancashire* (London: Ministry of Works: Ancient Monuments and Historic Buildings, 1943).

64   J.C. Dickinson, *Furness Abbey, Lancashire* (London: Department of the Environment: Ancient Monuments and Historic Buildings, 1965) – republished with amendments (London: English Heritage, 1987).

65   Joy, *A Regional History of the Railways of Great Britain*, p. 130; K.J. Norman, *The Furness Railway, 1: The Line Described* (Kettering: Silver Link Publishing, 2001), pp. 25–6. The remaining wing, which used to house the station's refreshment rooms, was converted to the Abbey Tavern. The site of the hotel is occupied by the present English Heritage car park and visitor centre, opened in 1982.

Chapter 12

The Post-Industrial Picturesque? Placing and Promoting Marginalised Millom

David Cooper

Wordsworth wrote:
'Remote from every taint of sordid industry'.
But you and I know better, Duddon lass.[1]

And maybe the ghost of Wordsworth, seeing further than I can,
Will stare from Duddon Bridge, along miles of sand and mud-flats
To a peninsula bare as it used to be, and, beyond, to a river
Flowing, untainted now, to a bleak, depopulated shore.[2]

Throughout his literary career, the work of the twentieth-century Cumbrian writer, Norman Nicholson (1914–1987), was consistently underpinned by an imaginative preoccupation with the poetics of place. In particular, Nicholson famously devoted much of his topographic writing – in both poetry and prose – to documenting the practice of everyday life in his native Millom: a townscape which, in the twenty-first century, is characterised by a complicated coalescence of post-industrial, coastal, estuarial, urban and even pastoral physical features and which, by extension, resists rigid topographic categorisation. The peripherality of this place, in purely positional terms, is indisputable as it is a built environment which is significantly, and problematically, detached from wider spatial networks. Even in the second decade of the twenty-first century, travelling to and from Millom is a relatively time-consuming process: it is an approximately 100-minute circuitous rail journey from the nearest mainline train station at Lancaster; and just under an hour is required to drive from the closest junction of the M6. This sequestration has a demonstrable impact upon the everyday spatial practices of the town's residents; and, concurrently, it is a physical geography which any visitor to Millom necessarily has to factor into his or her planned journey to the place.

At a localised level, Millom is a built environment which is individuated by spatial boundedness. To the south and east of the town lies the gritty expanse of

---

1 N. Nicholson, 'To the River Duddon', in *Five Rivers* (London: Faber and Faber, 1944), pp. 16–17 (p. 17). I am grateful to David Higham Associates for permission to quote Norman Nicholson-related material.

2 N. Nicholson, 'On the Dismantling of Millom Ironworks', in *Sea to the West* (London: Faber and Faber, 1981), pp. 49–50 (p. 50).

the tidal Duddon Estuary; and to the west is the Irish Sea. Immediately to the north-west is Black Combe; and, although the principal coastal road (A595) skirts beneath the distinctively rounded contours of the fell, the Combe dominates the local landscape thereby creating the illusion of a formidable psycho-geological barrier between Millom and the rest of industrial west Cumbria. The only direction in which spatial flows are not problematised by topographic features is to the north-east, as Millom is situated just beyond the south-western tip of the Duddon Valley (which ends as the river progresses into the Duddon Estuary just south of Broughton-in-Furness): a rustic space of 'cottages, farms, a church, fields, woods' clustered together within a 'narrow aisle of cultivated land running up between the fells'.[3] Yet, even in this direction, road travel seems to accentuate the place's physical and imaginative apartness. In driving to the Duddon village of Ulpha, the resident of Millom first has to journey out of the town on the A5093: a loop-road which connects Millom with the main A595 but which also serves to reinforce the community's peninsularity. Then, further north, the driver has to cross the river at the stone-arched Duddon Bridge: an architectural feature which historically marked the border between the old counties of Cumberland and Lancashire; and which, today, seemingly signifies socio-spatial difference at the same time as providing a link of communication between the pastoral and post-industrial communities.

Millom's peripherality, therefore, can be ascertained through the consultation of a map. The map-making process, however, necessarily involves the Cartesian reduction of the physical texturalities of geographical space to a series of fixed points upon a plane representational surface. Crucially, the two-dimensional map also fails to account for the ways in which spatial marginality can be reinforced – and perhaps even *produced* – by a concatenation of historical, cultural, social and economic factors. In *Places on the Margin: Alternative Geographies of Modernity*, Rob Shields proffers the uncontroversial assertion that marginality can be endowed upon a particular place as a direct result of its 'out-of-the-way geographic location'.[4] Shields immediately transcends the conceptual limitations of a purely locational understanding of geographic peripherality, however, to argue that a more complex sense of marginality can be created through a place's associations with 'illicit or disdained social activities', or its position as 'the Other pole to a great cultural centre'.[5] Shields thereby declares his interest in sites which are 'not necessarily on geographical peripheries' but which have clearly 'been *placed* [my italics] on the periphery of cultural systems of space in which places are ranked relative to each other'.[6]

---

3   N. Nicholson, *Greater Lakeland* (London: Robert Hale, 1969; repr. 1996), p. 53.

4   R. Shields, *Places on the Margin: Alternative Geographies of Modernity* (London: Routledge, 1991), p. 3.

5   Ibid.

6   Ibid.

Shields's monograph is structured around a series of case studies which illustrate how the socio-spatial construction of marginality can be underpinned by such transgressive tropes as liminality and the carnivalesque. On the surface, then, this nebulous concept of marginality would not appear to provide an appropriate theoretical framework for thinking about the Lake District: a landscape which seems to be uncomplicatedly emplaced within a cultural, if not geographic, centre. Since the development of Picturesque tourism in the second half of the eighteenth century, the Lakes has been consistently perceived, and projected, as a site of national importance; and this spatial role was most famously articulated when William Wordsworth declared, in his *Guide to the Lakes*, 'the Lakes in the North of England' to be 'a sort of national property, in which every man has a right and interest who has an eye to perceive and a heart to enjoy'.[7] Wordsworth's frequently polemical prose is destabilised by internal tensions and contradictions; but the profound influence of this key work of literary geography can be traced in the evolution of the environmental thinking which ultimately led to the formation of the Lake District National Park in the middle of the twentieth century.[8] The Lake District's prominent status within the national consciousness is now cemented and, as a result, the area can be defined as a culturally privileged 'leisure zone' within the seemingly 'homogenously uncultured industrial' space of the North.[9]

Yet, although Shields's 'cultural classification' of 'peripheral sites' might appear to be of limited value when applied to this 'high ranking' landscape, his exploration of the dialectical relationship between the centre and the margins carries the potential to open up thinking about the spatial relationships which are embedded *within* the cultural system of the Lake District.[10] That is to say, the imbricated concepts of geographic peripherality and cultural hierarchisation can provide appropriate theoretical foundations for considering the relationship between the physical and social geographies of the central Lakes and those difficult-to-classify topographies which lie on the fringes of, or just outside, the boundaries of the National Park: the flatlands of the Solway Plain, for instance; the estuarial spaces of the Furness Peninsula; the infrastructural spine of what may be described as the M6 corridor; and most saliently, within the context of this geo-specific case study, the scarred post-industrial 'edgelands' of west Cumbria.[11] The

---

7   W. Wordsworth, *A Guide Through the District of the Lakes*, in W. J. B. Owen and J. Worthington Smyser (eds), *The Prose Works of William Wordsworth*, 3 vols (Oxford: Oxford University Press, 1974), vol. 2, pp. 151–259 (p. 225).

8   See, for example, J. Bate, *Romantic Ecology: Wordsworth and the Environmental Tradition* (London: Routledge, 1991); S. Gill, *Wordsworth and the Victorians* (Oxford: Clarendon Press, 1998); and I. Whyte, 'William Wordsworth's *Guide to the Lakes* and the Geographical Tradition', *Area* 32 (2000), pp. 101–106.

9   Shields, *Places on the Margin*, p. 231.

10   Ibid., p. 3.

11   The term 'edgelands' can be used to denote the physical and abstract peripherality of West Cumbrian places within wider spatial networks. At the same time, however, the application of the noun to describe the difficult-to-define topographies of West Cumbria

perceived exceptionalism of the Lake District is founded upon the iconic contours of the landscape of the geographic centre. The terrains beyond the central dome of upland fells, though, are situated on the fringes of the dominant touristic narratives associated with this part of England; these edgelands remain on the geographic, imaginative and economic peripheries of that dominant spatial system. These locations are of a low ranking, therefore, within the wider context of what John Urry labels the 'place-myth' of the Lake District.[12]

In this essay I want to draw upon Shields's spatial thinking in order to explore the geographic, social, cultural and economic marginalisation of Millom; and, by extension, I want to map out some of the ways in which these mutually reinforcing processes of peripheralisation have impacted upon, and continue to shape, touristic practices in and around the town. To address this multi-layered marginalisation, the essay will be organised into three broadly chronological sections which trace key phases in Millom's spatial history (the pre-industrial, the industrial and the post-industrial) and which oscillate between consideration of the town as a locus for cultural representation, socio-economic development and tourism. The focus of this chapter, then, is site-specific as it examines how a complex interweaving of geography, history and socio-spatial processes have moulded the development of tourism in a singularly complicated place. At the same time, however, the aspiration is that the essay will also open up thinking regarding tourism practices in other 'out-of-the-way' places which have been subjected to analogous processes of 'symbolic exclusion'.[13]

Nineteenth-Century Millom: A Brief Spatial History

The relative remoteness of Millom immediately raises questions as to why and how this town, with a population of 7,132 at the 2001 census, originally came into being.[14] Until the middle of the nineteenth century, the area occupied by modern Millom was known as Millom Below and consisted of the village of Holborn Hill

---

corresponds, at least in part, with Marion Shoard's neologistic use of the term: a twenty-first-century form of landscape categorisation which has influenced the spatial thinking of the poets, Paul Farley and Michael Symmons Roberts, and which demands greater interrogation than is possible within the present context. See M. Shoard, 'Edgelands', in J. Jenkins (ed.), *Remaking the Landscape: the Changing Face of Britain* (London: Profile Books, 2002), pp. 117–46; and P. Farley and M. Symmons Roberts, *Edgelands: Journeys into England's True Wilderness* (London: Jonathan Cape, 2011).

12   J. Urry, *Consuming Places* (London: Routledge, 1995; repr. 2000), p. 195.

13   Shields, *Places on the Margin*, p. 5.

14   Office for National Statistics, 'Neighbourhood Statistics': http://www.neighbourhood.statistics.gov.uk/dissemination/LeadTableView.do;jsessionid=ac1f930c30d58c42a83b92cd406b9fe8ac18181e733e?a=7&b=793046&c=millom&d=16&e=15&g=432482&i=1001x1003x1004&m=0&r=1&s=1298550463722&enc=1&dsFamilyId=779&nsjs=true&nsck=true&nssvg=false&nswid=1276 (accessed 27 April 2011).

and a scattering of houses; and, although Millom Below had held a market charter since 1250, the community was essentially a rural village with a population, in 1841, of just 356.[15] Until the middle of the nineteenth century, therefore, everyday life in this part of Cumberland was characterised by a quiet pastoralism.

It was this quotidian ordinariness which attracted Wordsworth to the adjacent Duddon Valley at the beginning of the nineteenth century. In 1820, Wordsworth first published *The Duddon Sonnets*: a sequence of geo-specific poems in which the river is traced from its 'birthplace' at Wrynose, beneath Cockley Beck Bridge, alongside the settlements at Seathwaite and Ulpha, and out into the estuary at Foxfield.[16] By 1820 'the conversion of parts of the Lake District into a "literary landscape", which entailed the idealisation of a society and imagined way of life', was well under way; and the central Lakes was increasingly established as a site of cultural tourism.[17] In representing the Duddon Valley, therefore, Wordsworth self-consciously moved out and away from the central fells and towards one of the region's less familiar and culturally neglected sites: a valley which may have had profound imaginative significance for the Romantic poet but which was physically distant from the principal nexuses of touristic interest at Keswick and Ambleside.

To develop a fuller understanding of the geographic significance of Wordsworth's poetic project, it is necessary to re-emplace the sonnet sequence in its original publication context. As Stephen Gill explains, *The River Duddon, A Series of Sonnets: Vaudracour and Julia: And Other Poems. To which is Annexed, A Topographical Description of the Country of the Lakes, in the North of England* – to give the collection its full title – is 'a discrete volume' which 'is of the greatest historical significance', as Wordsworth uses the textual space to bring together verse and prose accounts of the Cumbrian topography thereby constructing a composite portrait of place.[18] The publication sets up what the textual critic, Neil Fraistat, would call a 'contextural poetic' as the sonnet sequence is accompanied by substantive supplementary notes and a version of the prose *Guide to the Lakes* unambiguously entitled 'Topographical Description of the Country of the Lakes'.[19] In the words of John Wyatt, this

---

15   A. Harris, *Cumberland Iron: The Story of Hodbarrow Mine 1855-1968* (Truro: D. Bradford Barton, 1970), p. 17.

16   W. Wordsworth, *The River Duddon. A Series of Sonnets*, in Geoffrey Jackson (ed.), *Sonnet Series and Itinerary Poems, 1820–1845* (Ithaca: Cornell University Press, 2004), pp. 56–75 (Sonnet I, p. 56, line 9).

17   C. O'Neill and J.K. Walton, 'Tourism and the Lake District: Social and Cultural Histories', in D.W.G. Hind and J.P. Mitchell (eds), *Sustainable Tourism in the English Lake District* (Sunderland: Business Education Publishers, 2004), pp. 20–47 (p. 20).

18   S. Gill, *William Wordsworth: A Life* (Oxford: Oxford University Press, 1989; repr. 1990), p. 333.

19   N. Fraistat, 'Introduction: The Place of the Book and the Book as Place', in Neil Fraistat (ed.), *Poems in their Place: The Intertextuality and Order of Poetic Collections* (Chapel Hill: The University of North Carolina Press, 1986), pp. 3–17 (p. 3). For more on the 'textual morphology' of Wordsworth's *Guide*, see Bate, *Romantic Ecology*, p. 44.

prose account provides a detailed geographical context, or 'authentication', for the geo-specific sonnet sequence which describes 'an unknown valley off the well-trodden route'.[20] The cross-reference to the authoritative prose text thereby assures the reader that this site is clearly located within the boundaries of the Lake District; it confirms the status of the Duddon as a location within the wider spatial system of the Lakes. Yet, simultaneously – and crucially – it is a site which remains at the outermost reaches of that cultural network.

Wordsworth uses his sonnet sequence to suggest that it is through its status as a site of geographic and cultural marginality that the Duddon Valley is emblematic of nationhood. As the spaces and places of the central Lakes were playing an increasingly prominent role within the national imagination, Wordsworth retreated to the margins and suggested that the imaginative value of the Duddon Valley resided in its condition as a site of unspoilt pastoralism which had remained resistant to potentially destructive socio-economic pressures: 'Child of the clouds! remote from every taint / Of sordid industry thy lot is cast'.[21] As James M. Garrett puts it: 'the writing of the local serves to exemplify what Wordsworth feels must be preserved and what he fears is fast ebbing away in the face of a nationalistic assault emanating from London'.[22] This anti-metropolitan reading of the text is reinforced by Wordsworth's reference to the 'sovereign Thames' in the sonnet in which the 'Majestic Duddon' is shown to expand 'over smooth flat sands /Gliding in silence with unfettered sweep' as it moves 'in radiant progress toward the deep' of the Irish Sea: 'Beneath an ampler sky a region wide /Is opened round him: – hamlets, towers, and towns, /And blue-topped hills, behold him from afar'.[23] Wordsworth uses the River Duddon, then, to construct a spatial model of positive peripherality.

The topo-authenticity – and, by extension, the imaginative currency – of Wordsworth's pastoral poem of nationhood was to be radically problematised, however, in the middle of the nineteenth century as the relatively unremarkable history of the southerly reaches of the Duddon was to be ruptured by the discovery of unprecedentedly rich deposits of haematite ore: a discovery which, in the appositely violent words of Nicholson, led to the town of Millom being 'yanked […] into being by a kind of geological Caesarian operation'.[24] In *Cumberland Iron: The Story of Hodbarrow Mine, 1855-1968*, A. Harris offers an exhaustive account of the industrial history of this corner of Cumbria; and he confirms that, at the time of the publication of Wordsworth's sonnet sequence in 1820, the peninsula, 'remained largely untouched by the forces of change which

---

20   J. Wyatt, *Wordsworth's Poems of Travel, 1819–42: 'Such Sweet Wayfaring'* (Basingstoke: Macmillan, 1999), p. 40.

21   Wordsworth, *The River Duddon*, II, p. 57, 1–2.

22   J.M. Garrett, *Wordsworth and the Writing of Nation* (Aldershot: Ashgate, 2008), p. 128.

23   Wordsworth, *The River Duddon*, XXXII, p. 74, 4–12.

24   N. Nicholson, *Wednesday Early Closing* (London: Faber and Faber, 1975), p. 175.

had already begun to transform the coal and iron districts to the north and the iron-mining across the Duddon'.[25] Over the course of the 1850s, however, there was an increasing amount of speculative interest in the area's telluric space; and, in 1856, a singularly thick body of iron ore was located at Hodbarrow – a mile to the south-east of modern-day Millom – with further significant discoveries being made in 1860 and 1868.[26] These discoveries resulted in the formation of the Hodbarrow Mining Company and the location's identity as a site of industry began to emerge. The mines at Hodbarrow expanded both rapidly and profitably; and, in 1867, two iron furnaces were built to the north of Hodbarrow on the edge of the present-day town (Figure 12.1). As a result of this expeditious industrial development, the local population grew exponentially as men expectantly arrived in south Cumberland from across Britain and Ireland, with a particular influx of workers from the mining fields of Cornwall.[27] By the time of the 1871 census, the population of Millom had multiplied to 2,656 and the gridded geometry of the town had begun to take shape.[28]

The second half of Millom's nineteenth century was defined by the continued expansion of the built environment which, in turn, was directly attributable to the ongoing productivity of the Hodbarrow mine: a narrative, in the words of J. D. Marshall and John K. Walton, 'of allied resourcefulness and good fortune, boldness, technical skill, and not a little concern with community building and the institutions of a flourishing small town'.[29] Within the spatial context of newly industrialised Millom, then, Wordsworth's topo-poetry was problematically – and perhaps even comically – anachronistic. As Garrett illustrates, Wordsworth's *The River Duddon* is founded upon the self-conscious celebration of the pastoral rootedness and rustic timelessness to be found in this corner of Cumbria.[30] By the end of the nineteenth century, however, Wordsworth's Duddon was flanked, as it glided out into estuarial space, by a landscape which had been thickened by the paraphernalia, practices and processes associated with the extraction of iron ore. As a result, the climax of *The River Duddon* now offered a nostalgic valorisation of an unrecognisably rural way of being for the *fin de siècle* reader emplaced on the Duddon peninsula.

---

25 Harris, *Cumberland Iron*, p. 17. Harris draws attention, however, to the existence of a small charcoal iron furnace at Duddon Bridge (p. 17): a material symbol of industry which is conspicuously absent in Wordsworth's portrait of the place but which has subsequently emerged as a key site of industrial archaeology. See J.D. Marshall and M. Davies-Shiel, *The Industrial Archaeology of the Lake Counties* (Beckermet: Michael Moon, 1977), pp. 42–4.

26 Harris, *Cumberland Iron*, pp. 21–4.

27 Ibid., p. 42.

28 Ibid., p. 74.

29 J.D. Marshall and J.K. Walton, *The Lake Counties from 1830 to the Mid-Twentieth Century* (Manchester: Manchester University Press, 1981), p. 51.

30 Garrett, *Wordsworth and the Writing of Nation*, pp. 126–48.

Iron Works, Millom.

Figure 12.1   1904 postcard of Millom Ironworks which were founded in 1867
Source: © Karen Pugh Collection, Millom Discovery Centre.

In summary, then, at the beginning of the nineteenth century this most southerly tip of Cumberland was considered too geographically remote to be incorporated within standard (post-) Picturesque tours of the region; and, for Wordsworth, the fact that the area sat outside dominant touristic narratives provided a source of positive peripherality. By the end of the 1800s, though, the emergence and growth of Millom as an iron-mining town meant that this particular built environment failed to correspond with the images of lakes and fells, tarns and crags, which understandably dominated post-Romantic touristic projections of the Lake District. Although the latter half of the nineteenth century was a period of economic growth and opportunity for those seeking employment in Millom, the construction of this functional townscape can be understood – within the context of regional tourism – as prompting a process of negative marginalisation in which the heavily industrialised tip of the Duddon peninsula remained ideologically excluded from the wider spatial system of the Lakes. Patently, the creation of a landscape of mine-shafts and foundations, railways sidings and terrace housing, placed Millom outside the cultural, and increasingly contested, space of 'Wordsworthshire': an aesthetically privileged topography in which, as Gill points out, late Victorian readers were necessarily negotiating the complex tensions embedded within the Romantic poet's (touristic) promotion of this pastoral region to develop proto-environmental thinking.[31] It was superfluous for the industrialised topography

---

31   E.S. Roberston, *Wordsworthshire* (London: Chatto and Windus, 1911).

of Millom to be incorporated within the poetry-fuelled debates regarding the management of the increasingly pressurised Lakeland landscape. The town was of little touristic interest; and, for those visitors entering the leisure zone of the Lakes – tourists travelling north on the coastal railway in order to access the western valleys – it was nothing more than a polluted other through which they had to pass. Ideologically, then, if not cartographically, Millom was placed beyond the boundaries of the Lake District map.

Placing the 'sub-rural' Town: Nicholson, the Lake District and Twentieth-Century Millom

As Harris illustrates, the apogee of Millom's industrial productivity, and economic prosperity, arrived either side of the turn of the twentieth century.[32] It would be an over-simplification to assert that the period after 1907 – the year of record mining production – was characterised by a trajectory of uniform decline as Harris's historical survey registers the subtle shifts and fluctuating fortunes which continued to take place over the course of the last century.[33] At the same time, however, Harris acknowledges that such 'interruptions were no more than temporary checks in a progression that was to be terminated only in 1968 by the closure of the mine'; and that the economic narrative of mining in twentieth-century Millom was 'the record of a slow but persistent decline'.[34] Within just one hundred years, therefore, this remote corner of Cumberland had been subjected to an accelerated process of landscape evolution: the area had been transformed from a site of agricultural pastoralism, to a densely populated and polluted urbanised landscape dominated by the architecture of the Hodbarrow mines and the Millom ironworks (which, inevitably, also closed in 1968), to a scarred site of post-industrialism in which the town's core *raison d'être* had seemingly ceased to be.

Perhaps predictably, the figurative and touristic marginalisation of Millom was subjected to administrative formalisation when, in 1949, the town was placed outside the boundaries of the new Lake District National Park which were legislatively encoded two years later. Understandably, the upper reaches of the Duddon Valley were included within the borders of this newly nationalised space; but the Victorian development of Millom as a site of industry led to the town's exclusion from the privileged leisure zone and the decision was made for the National Park boundary to go no further south than the Whicham Valley which skirts the southern flank of Black Combe. Millom – and its outlying settlements at Kirksanton, The Hill and The Green – had been politically ostracised from the wider cultural system of the Lakes.

---

32 Harris, *Cumberland Iron*, p. 117.
33 Ibid., p. 30.
34 Ibid., p. 111.

Nicholson, who spent almost all of his 73 years living at 14 St George's Terrace in the centre of Millom, dedicated much of his writing to the spatial history of the Lake District. The poet's topographic prose books are consistently underpinned by an ambition to draw the reader's attention to terrains beyond the all-too-familiar touristic route which takes visitors from Windermere to Ambleside to Grasmere and to Keswick. Moreover, Nicholson seeks to highlight the rich complexities of those spaces and places which are situated beyond the figurative walls of the National Park. It is a self-conscious opening up of the regional space which is visually encapsulated by the black-and-white photographs selected for inclusion in his debut topographic book, *Cumberland and Westmorland*: a publication which was issued at the historical moment at which the boundaries of the National Park were being mapped; and a publication which, saliently, contains images of Foxfield Sands, St Bees Head, Alston, Beetham and Burton-in-Kendal, alongside those touristic staples of Tarn Hows, the Langdale Pikes, Coniston Old Man and the Bowder Stone.[35]

At the beginning of *Cumberland and Westmorland*, Nicholson seeks to transcend the post-Romantic socialisation of landscape formally codified by the National Park by focusing upon telluric space.[36] For Nicholson, a sense of place is derived from the rock rather than the political or administrative organisation of the surface terrain; he then uses this preoccupation with geology to propose a spatial configuration of region which challenges the boundaries of the National Park which were about to come into being:

> The area which I call the Lake System is a large one, spreading a long way beyond the lakes. It is roughly an oval or ellipse standing on end. It apexes are Carlisle in the north and Carnforth in the south; its borders are the coast on the west, and on the east the valleys of the Eden and the Lune, or perhaps a line a little nearer the Yorkshire border.[37]

This spatial remodelling, built upon telluric unity rather than surface topography, collapses the post-Romantic hierarchisation of regional space. Nicholson prompts the reader to revisit the local map and directs his or her attention to ideologically marginalised sites; places, such as Millom, which have been situated outside the privileged, pastoral zone. Within this revised democratic configuration, Cumbria's industrial towns are positioned within the same spatial unit as the picture-postcard villages of the central Lakes; they are shown to be part of the same 'System'.

Nicholson endeavours to demonstrate the porosity between the two topographical categories through delineation of the spatial flows which govern

---

35 N. Nicholson, *Cumberland and Westmorland* (London: Robert Hale, 1949).

36 For an earlier exploration of these ideas see D. Cooper, 'The Poetics of Place and Space: Wordsworth, Norman Nicholson and the Lake District', *Literature Compass* 5 (2008), pp. 807–21.

37 Nicholson, *Cumberland and Westmorland*, p. 1.

everyday life; and he begins *Cumberland and Westmorland* with the explicit suggestion that there is a need to revaluate the idea – famously proposed by Wordsworth in his *Guide to the Lakes* – of a locatable centre to the Lake District:[38]

> For the society of the district, the hub is not a focal point but a point of departure. Two people living twenty miles apart or less on opposite sides of the hub might be a hundred miles apart for social purposes. [. . .] For practical and social purposes, the life of the dales does not leap-frog from one to another over the ridges, but flows up and down the valley.[39]

For Nicholson, the physical contours of the landscape lead to a heterotopic vision of space predicated upon everyday socio-economic practices. Nicholson points out that villages, or small towns, can be found at the head of each valley, including the built environments of Cleator Moor, Gosforth, Broughton-in-Furness and Ambleside. Beyond this circle, however, there exists 'an outer ring of market towns', such as Penrith, Appleby, Kendal, Ulverston, Whitehaven, Cockermouth, and [...] Wigton': sites at which 'real contact is made between the dales and the lowlands'.[40] Nicholson goes on to add further place-names to this list suggesting that, in spite of their industrial origins, 'a few towns like Workington, Millom and Barrow-in-Furness' serve a practical 'marketing' function – in the traditional sense of that term – as a result of their geographic locations.[41] This means that, for Nicholson, there is a quotidian connectedness between the town of Millom and the rural communities of the Duddon Valley; a spatial flow which cuts across the wandering boundary of the Lake District National Park. It is a direct relationship between town and country, therefore, in which Duddon Bridge is perceived to be a point of connectivity and through which Millom can be labelled a 'sub-rural' place.[42]

Nicholson's geographic obsession with spatial flows, processes and boundaries can be interpreted in several overlapping ways. In one sense, he is interested in documenting the practices which govern the everyday lives – including his own – of the residents of mid-twentieth-century Cumbria. At the same time, Nicholson's topographic writings can be seen to feed off, and back into, his overarching literary project to situate the authorial self in wider geographic and cultural contexts. Beyond this, though, he is simultaneously interested in analysing both historic and contemporaneous touristic habits through spatial frameworks: a textual practice which consistently informs what is arguably his most celebrated prose book,

---

38 Wordsworth's project to locate the 'centre' of the region results in the construction of an imaginary 'station', 'hanging midway' between 'Great Gavel, or Scawfell'. Wordsworth, *Guide to the Lakes*, p. 171.

39 Nicholson, *Cumberland and Westmorland*, p. 2.

40 Ibid., p. 3.

41 Ibid.

42 N. Nicholson, *Provincial Pleasures* (London: Robert Hale, 1959), p. 84.

The Lakers: The Adventures of the First Tourists.[43] The attempted recalibration of dominant spatial definitions of the Lakes emerges as a cardinal trope running through the poet's topographic writings as he questions the geographical, ideological and environmental wisdom of perceiving, and managing, the central Lakes as a privileged space: a preoccupation which corresponds with what he identifies to be the post-Picturesque touristic 'fashion [. . .] to concentrate on more and more intensive exploration of a continually shrinking area'.[44] By extension, Nicholson, in *Greater Lakeland*, uses his examination of everyday spatial flows as a foundation for re-thinking the relationship between the *touristic* centre and marginalised peripheries of the Lake District:

> The English Lake District is one of the most famous tourist centres in the world, yet the part of it which is most often visited is quite small – only the inner circle of the higher fells and the upper reaches of the dales. Outside that inner circle there is a far larger area, less known but containing a wide variety of countryside, much of it of great beauty, together with a wealth of social, historical and human interest greater, even, than that of the Lake District itself.[45]

Nicholson's overt ambition is to direct 'the visitor' away from 'the better-known inner parts' and towards 'these outlying parts of Greater Lakeland':[46] topographies which feature 'Windscale Atomic Station, the Marchon chemical factory at Whitehaven, Workington Steelworks, Barrow Docks and Carlisle Railway Depot' as well as the poet's home-town.[47] Writing from *within* a space which has been historically marginalised, Nicholson demands a cultural recalibration of the perceived relationship between the supposed centre and the outlying edgelands.

Marginality and Materiality: Tourism Practices in Twenty-First-Century Millom

Although Nicholson's polemical ambitions are transparent, the question remains as to what the twenty-first-century visitor might encounter if he or she follows the poet's advice and eschews the Lake District's principal tourist destinations to venture towards one of the region's more peripheral places. What efforts have been made to regenerate de-industrialised Millom since Nicholson's death in 1987? More particularly, in what ways have tourism professionals – at both county-wide and localised levels – endeavoured to repackage Millom as a post-

---

43   N. Nicholson, *The Lakers: The Adventures of the First Tourists* (London: Robert Hale, 1955).

44   Nicholson, *The Lakers*, p. 2.

45   Nicholson, *Greater Lakeland*, p. 11.

46   Ibid.

47   Ibid., p. 230.

industrial destination; and how do these tourism strategies and practices appear to intersect and interconnect with the sustained spatial thinking to be located within Nicholson's *oeuvre*?[48]

In 'Destination Cumbria, A 21st Century Experience', Cumbria Tourism articulate their 'Growth Marketing Strategy' for 2008–2012.[49] The executive summary consists of a series of bullet-pointed strategic aims, the first of which is to 'attract new/lapsed visitors through an awareness campaign with new messages and themes'.[50] Saliently, the second declared aim is to 'attract using the strongest attack brand "The Lake District" and disperse the benefits of tourism across Cumbria'.[51] This document, then, draws upon the dominant centrifugal model of the spatiality of the Lakes which can be traced back to Wordsworth's *Guide*. Crucially, however, the strategic vision transgresses the spatial limits imposed by the National Park boundaries to incorporate the traditionally marginalised landscapes to be found upon the fringes of Nicholson's 'Lake System' or 'Greater Lakeland'.

Cumbria Tourism's commitment to the strategy of 'attract and disperse' is manifested, in marketing terms, through the creation of the 'Western Lake District': a sub-brand of 'The Lake District, Cumbria', with its own promotional brochure and discrete website.[52] The textual content of the former encapsulates the organisation's overarching strategic intentions as the reader's attention is immediately drawn to some of the area's less obvious tourist destinations – including the 'bird-watching sites and nature reserves along the Solway Coast Area of Outstanding Natural Beauty', for instance, and 'the Egremont Castle illuminations' – as well as the more familiar site of 'Britain's Favourite View' at Wastwater.[53] The brochure also details 'a magnificent array of Events and Festivals'[54] and focuses on Cumbria's frequently forgotten status as a coastal county.[55] Yet, above all, four touristic features are foregrounded as the 'Western Lake District' is presented through a series of interlinking images: as a centre of award-winning locally produced food and drink; a site for (extreme) outdoors activity; a 'beautiful' and 'unspoilt' area which is defined by its topographical

---

48   I gratefully acknowledge the helpful information and professional insights enthusiastically volunteered by Mark Holroyd, South Copeland Tourism Officer, in the drafting of the final section of this chapter. At the same time, however, I must emphasise that the following commentary on contemporary tourist practices is entirely my own.

49   Cumbria Tourism, 'Destination Cumbria, A 21st Century Experience: Growth Marketing Strategy for Cumbria, 2008–2012' (Staveley, 2008).

50   Ibid., p. 3.

51   Ibid.

52   Cumbria Tourism, 'Western Lake District' (Staveley, 2009). Available at: http://www.western-lakedistrict.co.uk/ (accessed 27 April 2011).

53   Ibid., p. 3. Wastwater was named as the site of 'Britain's Favourite View' as part of a television series broadcast by ITV in 2007. See: http://www.itv.com/view/ (accessed 27 April 2011).

54   Ibid., p. 6.

55   Ibid., pp. 8–13.

heterogeneity; and a lived-space which is home to a series of 'warm' and welcoming communities.[56] The promotional literature constructs a bricolage out of these disparate strands to present an image of the 'Western Lake District' which is predicated upon a rhetoric of authenticity. That is to say, Cumbria Tourism's marketing strategy implicitly draws upon the historical peripheralisation of this area and incorporates this perceived marginalisation *within* the promotional discourse. The 'Western Lake District' is the home of 'true Cumbrian people' and distinctively 'authentic' food: communities and experiences which can only be accessed by those visitors who move beyond the central Lakes.[57]

Millom does not feature prominently within this positive reframing of the marginalised western fringes of the Lake District. The 'unspoilt' beach at Haverigg receives a cursory mention and is supplemented by a small image of Josefina de Vasconcellos's seven-tonne sculpture, 'Escape to Light', which looms above the sands. Millom also appears in a list of nine 'Western Lake District' towns on the back cover of the brochure.[58] Tellingly, Millom is described as a 'former fishing village' and the selected visual representation projects an essentially pastoral image of the town: the foreground of the photograph consists of grazing cattle; the middle-distance is dominated by the steeple of St George's Church; whilst the upper-third of the image consists of the backdrop provided by the lower slopes of Black Combe (Figure 12.2).[59] Beyond this carefully choreographed image, however, Millom does not figure in the promotional brochure; it does not appear to be integral to the 'attract and diverse' strategy into which the 'Western Lake District' promotional material clearly feeds. As a result, Millom continues to occupy a peripheral position within Cumbria Tourism's attempts to repackage this historically marginalised Cumbrian fringe.

Clearly, the fixity of Millom's position upon the map continues to problematise the efforts to attract visitors to a town which corresponds with what Frances Brown and Derek Hall define to be the essential basis for touristic 'peripherality': an area which 'suffers from *geographical* [my emphasis] isolation, being distant from core spheres of activity, with poor access to and from markets'.[60] Twenty-first-century Millom indubitably suffers, therefore, from its physical distance from other built environments promoted under the 'Western Lake District' banner: it is just over thirty miles south of the Georgian town of Whitehaven, for instance, which *does* feature prominently within the tourist literature and which also serves as the centre of political and administrative power for the Borough Council (Copeland) in which Millom is situated. Yet this touristic placement of Millom on the margin of the margins can also be understood to have been produced by an ideological process

---

56   Ibid., pp. 3–4.

57   Ibid., p. 4.

58   Ibid., p. 24.

59   Ibid.

60   F. Brown and D. Hall, 'The Paradox of Peripherality', in F. Brown and D. Hall (eds), *Tourism in Peripheral Areas: Case Studies* (Clevedon: Channel View Publications, 2000), pp. 1–6 (p. 2).

Figure 12.2 'Sitting by the Duddon River Estuary with Black Combe Fell on its doorstep, the former fishing village of Millom is a small coastal town which has a varied history'
Source: © Brian Sherwen, courtesy of Cumbria Tourism.

to comply with not only the visitor's imaginative preconceptions of the landscape of the central Lakes but also the new touristic narratives attached to the rebranded 'Western Lake District'. The perceived difficulties of situating the town within wider spatial networks may have provided Nicholson with richly problematic material for both lyrical poetry and frequently polemical prose; but its geographic remoteness and industrial history continue to underpin its peripheralisation within the wider frameworks of twenty-first-century tourism strategies and policies.

Yet, in spite of this ongoing marginalisation, there are more localised examples of tourism practices which are founded upon a self-conscious preparedness to engage with the complex singularity of this place. What is more, the grassroots projects to attract greater visitors to Millom *do* correspond with the key 'thematic campaigns' which Cumbria Tourism confidently predict will 'appeal to [...] specific target markets and offer the quality experience that they are looking for'.[61] The first of these overarching 'thematic' preoccupations is 'culture': an appositely malleable umbrella term which is used by Cumbria Tourism to cover 'traditional history, heritage, gardens and literature, as well as contemporary art, festivals and events'.[62] A communal enthusiasm to document, celebrate and communicate Millom's

---

61 'Destination Cumbria', p. 9.
62 Ibid.

layered history is most obviously, and conventionally, manifested by the Millom Discovery Centre: an accredited museum which is located at the town's railway station and whose displays include detailed information about the area's industrial past.[63] It is also materially evident in Colin Telfer's statue, 'The Scutcher', which commemorates the town's industrial past and which is located in the Market Square.

The fluidity of Cumbria Tourism's definition of culture, however, also allows for consideration of the way in which the actual sites of industry are now primarily perceived and managed as 'green' spaces of leisure within the urbanised infrastructure of the town. The site once occupied by Millom Ironworks is now a surreally lunar landscape of large slag piles around which information boards helpfully summarise the history of this particular plot of land. What is more, a panoramic view-point board furnishes the visitor with the names of the fells that appear on the 'Duddon Skyline' as he or she looks back up the estuary – a vista which swings from Black Combe in the north to Kirkby Moor in the south and which takes in the Scafell Range, Coniston Old Man and, on a clear day, High Street – and provides the full text of Nicholson's late poem, 'Scafell Pike' (Figure 12.3).[64] In contrast to other worked sites in the 'Western Lake District' – the Haig Colliery Mining Museum on the outskirts of Whitehaven and the Florence Mine Heritage Centre at Egremont (which closed in 2007) – there is no great material presence to signify the industrial past or for the archeologically minded visitor to explore. As a result, the open space – situated on a raised plateau at the end of a dead-end street on the edge of the town – is characterised, in part, by a haunting absence. For the poet and topographer, Neil Curry, the absence of things at this location is historically and socially problematic: 'It [the landscape of industrialism] was all quickly shovelled away as if it were something to be ashamed of'.[65]

Crucially, though, the site is not, to draw upon Tim Edensor's work on industrial ruins, a valueless space which is 'saturated with negativity'; nor, in spite of surface appearances, is it a zone free of life and energetic movement.[66] Rather, the site is now managed by a Local Nature Reserve committee and the slag piles are slowly being colonised by a rich range of rare flora and fauna, including orchids and yellow wort; most famously, and most noisily, the site also provides a dwelling-place for a community of natterjack toads. A second nature reserve, owned and managed by the Royal Society for the Protection of Birds, can be found at nearby Hodbarrow and provides a home for, amongst other species, great crested grebes,

---

63 In 2010, Millom Folk Museum was rebadged as Millom Heritage Museum and Visitor Centre; and, just two years later, the name Millom Discovery Centre was introduced. These nomenclatural shifts – which can be clearly situated in wider debates within the museums sector – illustrate the tensions inherent in ensuring that the presentation of local histories appears sufficiently dynamic to attract visitors from outside that community.

64 Nicholson, 'Scafell Pike', in *Sea to the West*, pp. 11–12.

65 N. Curry, *The Cumberland Coast* (Carlisle: Bookcase, 2007), p. 181.

66 T. Edensor, *Industrial Ruins: Space, Aesthetics and Materiality* (Oxford: Berg, 2005), p. 7.

Figure 12.3   The Millom Ironworks Nature Reserve against the backdrop of Black Combe.

Note: The panoramic view-point board, which looks out across the Duddon Estuary, is placed on the slate plinth seen in the middle-distance

Source: © The author.

terns and red-breasted mergansers, as well as more natterjacks. Here, on the edge of the Duddon Estuary, the built infrastructure of industry has been manipulated for post-industrial purposes: the Outer Sea Wall, originally constructed to maximise mining productivity, has been used to create an artificial lagoon for waterfowl and waders; and visitors can observe this protectively bounded space from a hide located on the Sea Wall itself (Figure 12.4). The imbrication of the man-made and the organic to be located at these two post-industrial sites corresponds with what the nature writer, Richard Mabey, describes as the 'unofficial countryside' of 'sites inside an urban area' which provide the 'right living conditions for some plant or creature': 'the water inside abandoned docks and in artificially created reservoirs; canal towpaths, and the dry banks of railway cuttings; allotments, parks, golf courses and gardens and so on'.[67] Together, then, these twin nature reserves

---

67   R. Mabey, *The Unofficial Countryside* (Wimborne Minster: Little Toller Books, 2010), p. 20. Mabey's text, which is an acknowledged influence on the topographical concept of edgelandness articulated by Farley and Symmons Roberts, was first issued in 1973.

invite the first-time visitor to recalibrate his or her touristic preconceptions of the Cumbrian landscape and, by extension, to test new ways of understanding and conceptualising place.

Figure 12.4   The lagoon at the RSPB Nature Reserve, Hodbarrow
Source: © The author.

By extension, these sites also prompt new ways of phenomenologically experiencing place. Understandably, the tourism history of the Lake District has been predicated upon what the cultural geographer, Denis Cosgrove, calls the 'geographies of vision'.[68] This emphasis on 'the ocular act of registering the external world' began, of course, with the landscape enframing strategies promoted by the Picturesque; and, although the Romantic poets moved towards a phenomenological articulation of what it means to be-in-the-world, the history of cultural tourism in the area has been dominated by the exploration and documentation of ways of seeing the lakes and fells.[69] It is the dominant form of landscape engagement to which the panoramic view-point at Millom Ironworks nature reserve demonstrably caters as the 'tourist gaze' is directed away from the immediate environment and towards

68   D. Cosgrove, *Geography and Vision: Seeing, Imagining and Representing the World* (London: I.B. Tauris, 2008).

69   Ibid., p. 5.

the uplands of the central fells.[70] Saliently, however, the walk across the cratered landscape once occupied by the ironworks forces the visitor to look down at his or her feet and, by extension, to engage with the flora and fauna emerging out of the slag. The embodied experience of moving across this edgeland, then, physically forces the visitor to combine the topographical distancing strategies inextricably associated with (post-) Picturesque viewing practices with a microscopic focus on the immediate and the near-at-hand.

Nicholson self-consciously drew upon the cultural privileging of seeing when, in 1969, he described the coastal landscape between Whitehaven and Workington as an example of 'the industrial picturesque':[71] a label which might have been equally applicable to the 'romantic landscape of caved-in shafts, subsidences, abandoned railway lines, broken-down chimneys and pit-gear' which he recorded at Hodbarrow.[72] Many of the material objects which Nicholson archived in the late 1960s, however, are no longer in place at post-industrial Hodbarrow. Yet, crucially, this absence of industrial things problematises the possibility of an unthinkingly visual engagement with environment and prompts the visitor to reflect further on his or her experience of being-in-place. Although this is a site which may be characterised by a haunting absence, it is also a complicatedly living, and biologically diverse, post-industrial landscape of smells and sounds as well as sights; a site which invites the visitor to adopt a multi-sensory openness to environment. The post-industrial nature reserves, then, collectively serve to open up what might be described as an eco-tourism of embodiment.

This emphasis on the body can also be traced in the second of Cumbria Tourism's 'thematic' priorities: 'outdoor activities' and the declared 'ambition for the destination to be recognised as Adventure Capital UK'.[73] Once again, localised tourism projects in Millom can be seen as feeding off, and back into, this county-wide strategy exemplified, for instance, by ongoing attempts to improve cycle access in the area and the use of the lagoon at Hodbarrow for water-sports. Moreover, Millom is also providing the base for the annual 'Black Combe Walking Festival': a series of guided walks and talks taking place in and around the Duddon peninsula each May and June.[74] The area, therefore, is being repackaged as a space for physical activity; it is being promoted as an unbounded space in which visitors can do as well as look.

There are other ways in which Millom's tourism professionals have been seeking to attract greater visitors to the area. There have been attempts, for instance, to benefit from the third of Cumbria Tourism's 'Thematic Campaigns' – an emphasis on 'Food and Drink' which has been promoted under the 'Taste District Cumbria'

---

70   J. Urry, *The Tourist Gaze: Leisure and Travel in Contemporary Societies* (London: Sage, 1990).

71   Nicholson, *Greater Lakeland*, p. 178.

72   Ibid., p. 129.

73   'Destination Cumbria', p. 9.

74   South Copeland Tourism Community Interest Company, 'Black Combe Walking Festival': http://www.walkingfest.org.uk/index.html (accessed 27 April 2011).

brand – through a celebration of local produce in a 'Beer "n" Bangers Festival' in 'the South Western Lake District' (including Millom).[75] The formation of a Norman Nicholson Society has also led to the town providing the natural centre for a series of literary events: readings, talks and participatory workshops which focus on the examination of Nicholson's work and, which simultaneously, use those writings to explore wider cultural and geographic issues. Alongside this, the attic room in Nicholson's former home – the ground floor of which is presently occupied by the Nicholson House Coffee Shop – is annually opened to the public as part of the town's programme of events for the 'Heritage Open Days' scheme.[76] Perhaps most strikingly, a contrast to the sense of industrial absence experienced at the sites of Millom Ironworks and Hodbarrow Mine is offered at Millom Rock Park: a private site, to the north-east of the town, owned and managed by Aggregate Industries, in which visitors are invited to develop their telluric understanding through both a series of geological samples and interpretative displays ('Rock Street') and by gazing 'into the depths of Ghyll Scaur' which remains a working quarry.[77]

Conclusion

The interrelated factors of geographic remoteness and less-than-straightforward accessibility continue to problematise the attempts to attract tourists to Millom: challenges which are unhelpfully exacerbated by the fact that, at present, no train services operate in or out of the town on Sundays; logistical difficulties which fuel ongoing debates regarding the need for a new road bridge linking Millom and Askam across the Duddon Estuary. Connected with this, the business of attracting visitors to the Duddon peninsula is also currently hindered by the relative lack of accommodation providers. Yet, in spite of these infrastructural problems, recent tourism projects centred on Millom have been characterised by a uniform desire to work with, rather than against, the complex history of the town; projects which have been predicated upon local, on-the-ground place-making, rather than the transplantation of a homogenous county-wide strategy. Through these projects, the problem of peripherality has been positively reframed. That is to say, the concept of Millom's marginality has been appropriated, at a community level, and redefined from within: a process which satisfies Chris Murray's demand that any

---

75   South Copeland Tourism Community Interest Company, 'Beer "n" Bangers Festival': http://www.beernbangers.com (accessed 27 April 2011). It is salient that the formal name of this company signifies the ongoing attempts to imbricate tourism activity with the concerns of the local residents.

76   English Heritage, 'Heritage Open Days': http://www.heritageopendays.org.uk/ (accessed 27 April 2011).

77   Bardon Aggregates, 'Millom Rock Park': http://www.millomrockpark.org.uk/ html/look_into_the_volcano.html (accessed 27 April 2011).

authentic 'place-marketing programme' must be founded upon sensitivity to the particularities and singularities of a location.[78]

These self-conscious place-making strategies are exemplified by a promotional leaflet for 'Millom and Haverigg' which has been printed and distributed by The Black Combe and Duddon Estuary Tourism Group: a leaflet in which text and images are brought together to showcase Millom as both a site of geographic remoteness and tranquillity and a landscape which has been sculpted by the demands of heavy industry.[79] Saliently, this leaflet also draws attention to Millom's status as a 'Hidden Britain Centre': a charity-run project, organised in partnership with enjoyEngland, designed to promote 'places that are too special to remain hidden – yet too precious to spoil'.[80] It is possible to critique the 'Hidden Britain' scheme as yet another attempt to facilitate the 'discovery' of untrodden wildernesses which have been omitted from the touristic map; and this spatial categorisation clearly complies with what Brown and Hall label the 'paradox of peripherality'.[81] Yet the accompanying text on the local leaflet serves to encapsulate many of the intersecting, and occasionally conflicting, issues which I have highlighted over the course of this essay:

> Hidden Britain Centres help you discover, explore and become part of, the fascinating wealth of local landscape, history, culture, food and community life in undiscovered parts of Cumbria, something that is rarely possible in better known holiday areas, providing a truly different and memorable experience.[82]

On one level, this quotation foregrounds the key promotional strategies which have been prioritised by Cumbria Tourism and which have filtered down to Millom. Yet, significantly, the text transcends the promotion of local topography, heritage and cuisine, to intimate that it might also be possible for the visitor to immerse him or herself within 'community life' in this out-of-the-way place; and the suggestion that the tourist may 'become part of' a place even implies the potential for a phenomenological integration into a particular localised way of being. It is also significant that the leaflet draws upon the culturally entrenched dualistic model of Cumbria. Crucially, though, the text reverses the established ideological relationship between the centre and the periphery as Millom is repackaged as offering the tourist an authentic experience of Cumbrian life which is founded upon the status of the town as a lived-space rather than as a site of leisure; the town is presented as the site of everyday, and embodied, spatial practices rather than

---

78   C. Murray, *Making Sense of Place: New Approaches to Place Marketing* (Stroud: Comedia, 2001), p. 111.

79   Black Combe and Duddon Estuary Tourism Group, 'Millom and Haverigg: A Poet's Land of Sand, Mines and Mountains'.

80   'Hidden Britain': http://www.hidden-britain.co.uk/ (accessed 27 April 2011).

81   Brown and Hall, 'The Paradox of Peripherality'.

82   'Millom and Haverigg'.

the subject of a distanced tourist gaze. In other words, the 'placial' margins are privileged over what might be described as the non-place of the central Lakes.[83] Ultimately, then, the spatial characteristics projected in this leaflet chime with the sustained geographic thinking of Millom's most celebrated archivist, Norman Nicholson. Both Nicholson's topographic writings, and contemporary tourist projects and discourses, prompt further thinking on what it means to situate Millom as a place on the margins of the Lake District; and, crucially, such questions are being asked from inside this site of geographic and cultural outsideness.

---

83   E.S. Casey coins the neologism, 'placial', to denote a situated, embodied being-in-the-world which is contrasted to the totalising abstraction of landscape and environment suggested by 'spatial'. E.S. Casey, *Representing Place: Landscape Painting & Maps* (Minneapolis: University of Minnesota Press, 2002), p. 351.

Select Bibliography

Bate, J. *Romantic Ecology: Wordsworth and the environmental tradition*. London: Routledge, 1991.

Bicknell, P. and Woof, R. (eds) *The Discovery of the Lake District 1750–1810: a context for Wordsworth*. Grasmere: Trustees of Dove Cottage, 1982.

Bicknell, P. and Woof, R. (eds) *The Lake District Discovered 1810–1850: the artists, the tourists, and Wordsworth*. Grasmere: Trustees of Dove Cottage, 1983.

Bouch, C.M.L. and Jones, G.P. *The Lake Counties 1500–1830: a social and economic history*. Manchester: Manchester University Press, 1961.

Bowden, M. (ed.) *Furness Iron: the physical remains of the iron industry and related woodland industries of Furness and southern Lakeland*. Swindon: English Heritage, 2000.

Brodie, I.O. *Forestry in the Lake District*. Kendal: Friends of the Lake District, 2004.

Brown, G. *Herdwicks: Herdwick sheep and the Lake District*. Kirkby Stephen: Hayloft, 2009.

Brown, M.E. *A Man of No Taste Whatsoever: Joseph Pocklington 1736–1817*. Milton Keynes: AuthorHouse, 2010.

Brunskill, R.W. *Traditional Buildings of Cumbria: the county of the Lakes*. London: Cassell, 2002.

Buzard, J. *The Beaten Track: European tourism, literature, and the ways to 'culture', 1800–1918*. Oxford: Clarendon Press, 1998.

Cannadine, D. *G.M. Trevelyan: a life in history*. London: Harper Collins, 1992.

Capstick, M. *Patterns of Rural Development: a study of North Westmorland and its problems*. Kendal: Westmorland County Council, 1970.

Caseby, R. *The Opium-Eating Editor*. Kendal: Westmorland Gazette, 1985.

Collingwood, W.G. *Lake District History*. Kendal: Titus Wilson, 1928.

Cooper, D. 'The Poetics of Place and Space: Wordsworth, Norman Nicholson and the Lake District', *Literature Compass* 5 (2008): 807–21.

Cosgrove, D. *Geography and Vision: seeing, imagining and representing the world*. London: I.B. Tauris, 2008.

Crist, E. 'Against the Social Construction of Nature and Wilderness', *Environmental Ethics* 26 (2004): 4–24.

Crook, J.M. 'Privilege and the Picturesque: new money in the Lake District, 1774–1914', in *The Rise of the* Nouveaux Riches: *style and status in Victorian and Edwardian architecture*. London: J. Murray, 1999: 79–100.

Dade-Robertson, C. *Furness Abbey: romance, scholarship and culture*. Lancaster: Centre for North-West Regional Studies, 2000.

Darby, W.J. *Landscape and Identity: geographies of nation and class in England*. Oxford: Berg, 2000.

Dawson, J. and Briggs, D. *Wordsworth's Duddon Revisited*. Milnthorpe: Cicerone Press, 1988.

De Quincey, T. *Recollections of the Lake Poets*. Edinburgh: Adam & Charles Black, 1862.

Denyer, S. *Traditional Buildings and Life in the Lake District*. London: Victor Gollancz, 1991.

Denyer, S. *Lake District Landscapes*. London: National Trust, 1994.

Denyer, S. *Beatrix Potter at Home in the Lake District*. London: Frances Lincoln, 2000.

Edensor, T. *Industrial Ruins: space, aesthetics and materiality*. Oxford: Berg, 2005.

Frey, B.S., Pamini, P. and Steiner, L. *What Determines the World Heritage List? An econometric analysis*. Zurich: University of Zurich, Department of Economics, Working Paper No. 1, January 2011.

Garrett, J.M. *Wordsworth and the Writing of Nation*. Aldershot: Ashgate, 2008.

Gray, T. *Thomas Gray's Journal of his Visit to the Lake District in October 1769*, edited by W. Roberts. Liverpool: Liverpool University Press, 2001.

Griffin, A.H. *The Coniston Tigers: seventy years of mountain adventure*. Wilmslow: Sigma Leisure, 2000.

Hall, B. *The Royal Windermere Yacht Club 1860–1960*. Altrincham: John Sherratt & Son, 1960.

Hall, M. (ed.) *Towards World Heritage: international origins of the preservation movement, 1870–1930*. Farnham: Ashgate, 2011.

Hall, M. 'The Politics of Collecting: the early aspirations of the National Trust, 1883–1913', *Transactions of the Royal Historical Society*, 6th series, 13 (2003): 345–57.

Hankinson, A. *A Century on the Crags: the story of rock climbing in the Lake District*. London: J.M. Dent & Sons, 1988.

Hankinson, A. *The Regatta Men*. Milnthorpe: Cicerone Press, 1988.

Hanley, K. 'Wordsworth's Region of the Peaceful Soul', in *From Lancaster to the Lakes, the Region in Literature*, edited by K. Hanley and A. Milbank. Lancaster: Centre for North-West Regional Studies, 1992.

Hanley, K. and Walton, J.K. *The Construction of Cultural Tourism: John Ruskin and the tourist gaze*. Bristol: Channel View, 2010.

Hansen, P.H. 'British mountaineering, 1850–1914', Ph.D. thesis, Harvard University, 1991.

Harman, P.M. *The Culture of Nature in Britain, 1680–1860*. New Haven & London: Yale University Press, 2009.

Harris, A. *Cumberland Iron: The story of Hodbarrow Mine 1855–1968*. Truro: D. Bradford Barton, 1970.

Healey, J. 'Agrarian Social Structure in the Central Lake District, c. 1574–1830: the fall of the "Mountain Republic"?', *Northern History*, 44, 2 (2007): 73–91.

Hillman, A. 'Common rights to stone, peat and coal in South Westmorland before and after enclosure', Ph.D. thesis, Leeds Metropolitan University, 2011.

Hyde, M., Pevsner, N. et al. *The Buildings of England: Cumbria*. New Haven & London: Yale University Press, 2010.

Irvine, W. and Anderson, A.R. 'The impact of foot and mouth disease on a peripheral tourism area' (2005): https://openair/rgu.ac.uk/bitstream/10059/214/1/Anderson17.pdf.

Jackson, H. and Jackson, M. *Lakeland's Pioneer Rock-Climbers: based on the visitors book of the Tysons of Wasdale Head, 1876–1886*. Clapham, North Yorkshire: Dalesman Books, 1980.

Jones, O.G., *Rock Climbing in the English Lake District*, 2nd edition. Didsbury: E.J. Morten, 1973.

Joy, D. *The Lake Counties. Regional History of the Railways of Great Britain 14*, 2nd edition. Newton Abbot: David St John Thomas, 1990.

Linder, L. (ed.) *The Journal of Beatrix Potter from 1881 to 1897*. London & New York: F. Warne, 1966.

McCormick, T. 'Wordsworth, hill farming and the making of a cultural landscape', *North West Upland Farming* (2009): http://www.cumbriahillfarming.org.uk/hillfarming/wordsworth.html.

McCracken, D. *Wordsworth and the Lake District: a guide to poems and their places*. Oxford: Oxford University Press, 1984.

McGhie, L. 'Consumer and consumption, 1650–1750: a study of household goods and the middling sort in South Westmorland and Furness', M.Phil. thesis, University of Central Lancashire, 2002.

Marshall, J.D. *Furness and the Industrial Revolution*. Barrow-in-Furness: Barrow-in-Furness Library and Museum Committee, 1958.

Marshall, J.D. *Old Lakeland: some Cumbrian social history*. Newton Abbot: David & Charles, 1971.

Marshall, J.D. '"Statesmen" in Cumbria: the vicissitudes of an expression', *Cumberland and Westmorland Archaeological and Antiquarian Society Transactions*, 2nd series, 72 (1972): 249– 83.

Marshall, J.D. 'Agrarian Wealth and the Social Structure in Pre-industrial Cumbria', *Economic History Review* 33 (1980): 503–21.

Marshall, J.D. 'Out of Wedlock: perceptions of a Cumbrian social problem in the Victorian context', *Northern History* 31 (1995): 194–207.

Marshall, J.D. and Davies-Shiel, M. *The Industrial Archaeology of the Lake Counties*. Beckermet: Michael Moon, 1977.

Marshall, J.D. and Walton, J.K. *The Lake Counties from 1830 to the Mid-twentieth Century*. Manchester: Manchester University Press, 1981.

Martineau, H. *A Complete Guide to the Lake District of England*, 4th edition. Windermere: J. Garnett & London: Simpkin, Marshall & Co., no date [1871].

Matless, D. 'Moral Geographies of English Landscape', *Landscape Research* 22 (1997): 141–55.

Matless, D. *Landscape and Englishness*. London: Reaktion, 1998.

Menuge, A. 'Belle Isle and the Invention of the Lake District', in *Town and Country: contemporary issues at the rural/urban interface*, edited by M.F. Hopkinson. York: PLACE Research Centre, 2000: 71–82.

Menuge, A. 'Patriotic Pleasures: boathouses and boating in the English Lakes', in *Living, Leisure and Law: eight building types in England 1800–1914,* edited by Geoff Brandwood. Reading: Spire Books, 2010: 53–72.

Murdoch, J. (ed.) *The Discovery of the Lake District: a northern Arcadia and its uses*. London: Victoria & Albert Museum, 1984.

Murdoch, J. (ed.) *The Lake District: a sort of national property. Papers presented to a symposium held at the Victoria & Albert Museum, 20–22 October 1984.* Cheltenham: Countryside Commission & London: Victoria & Albert Museum, 1986.

Nicholson, N. *Five Rivers.* London: Faber & Faber, 1944.

Nicholson, N. *The Lakers: The Adventures of the First Tourists.* London: Hale, 1955.

Nicholson, N. *Provincial Pleasures.* London: Robert Hale, 1959.

Nicholson, N. *Greater Lakeland.* London: Robert Hale, 1969.

O'Neill, C. 'Visions of Lakeland: tourism, preservation and the development of the Lake District 1919–1939'. Ph.D. thesis, Lancaster University, 2000.

O'Neill, C. and Walton, J.K. 'Tourism in the Lake District: social and cultural histories', in *Sustainable Tourism in the English Lake District*, edited by D.W.G. Hind and J.P. Mitchell. Sunderland: Business Education Publishers, 2004: 20–47.

Pace, J. and Scott, M. (eds) *Wordsworth in American Literary Culture.* Basingstoke: Palgrave Macmillan, 2005.

Powell, C. and Hebron, S. *Savage Grandeur and Noblest Thoughts: discovering the Lake District 1750–1820.* Grasmere: Wordsworth Trust, 2010.

Rawnsley, E. *Canon Rawnsley: an account of his life.* Glasgow: Maclehose, Jackson and Co., 1923.

Rawnsley, H.D. *Literary Associations of the English Lakes,* 2 vols. Glasgow: J. MacLehose, 1893.

Ritvo, H. *The Dawn of Green: Manchester, Thirlmere and modern environmentalism*. Chicago: University of Chicago Press, 2009.

Rodgers, C.P., Straughton, E.A., Winchester, A.J.L. and Pieraccini, M. *Contested Common Land: environmental governance past and present.* London: Earthscan, 2001.

Rollinson, W. 'William Gershom Collingwood, 1854–1932', in revised edition of *The Lake Counties* by W.G. Collingwood. London: J.M. Dent & Sons, 1988.

Ruskin, J. *Iteriad; or three weeks among the Lakes*, edited by J.S. Dearden. Newcastle upon Tyne: Frank Graham, 1969.

Searle, C.E. 'Custom, Class Conflict and Agrarian Capitalism: the Cumbrian customary economy in the eighteenth century', *Past and Present* 110 (1986): 106–33.

Selincourt E. de (ed.) *The Letters of William and Dorothy Wordsworth*, 2nd edition, revised by C.L Shaver, M. Moorman and A.G. Hill, 8 vols. Oxford: Clarendon Press, 1967–93.

Snape, R. 'The Co-operative Holidays Association and the Cultural Formation of Countryside Leisure Practice', *Leisure Studies* 23 (2004): 143–58.

Symonds, H.H. *Afforestation in the Lake District*. London: Dent, 1936.

Taylor, H. *A Claim on the Countryside*. Edinburgh: Keele University Press, 1997.

Thompson, B.L. *The Lake District and the National Trust*. Kendal: Titus Wilson, 1946.

Thompson, T.W. *Wordsworth's Hawkshead*. Oxford: Oxford University Press, 1970.

Tolia-Kelly, D.P. 'Fear in Paradise: the affective registers of the English Lake District landscape re-visited', *The Senses and Society* 2 (2007): 329–51.

Townend, M. *The Vikings and Victorian Lakeland: the Norse medievalism of W.G. Collingwood and his contemporaries*. Kendal: Titus Wilson, 2009.

Walton, J.K. 'Canon Rawnsley and the English Lake District', *Armitt Library Journal* 1 (1998): 1–17.

Walton, J.K. 'National Parks and Rural Identities: the North York Moors', in *Rural Life in the Nineteenth and Twentieth Centuries: regional perspectives*, edited by M. Tebbutt. Manchester: Conference of Regional and Local Historians, 2004: 115–31.

Walton, J.K. and McGloin, P.R. 'The Tourist Trade in Victorian Lakeland', *Northern History* 17 (1981): 152–82.

Welberry, K., 'Arthur Ransome and the Conservation of the English Lakes', in *Wild Things: children's culture and ecocriticism*, edited by S.L. Dobrin and K.B. Kidd. Detroit: Wayne State University Press, 2004.

West, T. *A Guide to the Lakes: dedicated to the lovers of landscape studies, and to all who have visited, or intend to visit the Lakes in Cumberland, Westmorland and Lancashire. By the author of the Antiquities of Furness*. London: Richardson & Urquhart and Kendal: W. Pennington, 1778.

Westall, O.M. (ed.) *Windermere in the Nineteenth Century*, 2nd edition. Lancaster: Centre for North-West Regional Studies, 1991.

Westaway, J.H. 'The German Community in Manchester, Middle-Class Culture and the Development of Mountaineering in Britain, c. 1850–1914', *English Historical Review* 124 (2009): 571–604.

Wheeler, M. (ed.) *Ruskin and Environment: the storm cloud of the nineteenth century*. Manchester: Manchester University Press, 1995.

Whyte, I. 'Wordsworth's *Guide to the Lakes* and the Geographical Tradition', *Armitt Library Journal* 1 (1998): 18–37.

Whyte, I. 'William Wordsworth's *Guide to the Lakes* and the Geographical Tradition', *Area* 32 (2000): 101–06.

Whyte, I. *Transforming Fell and Valley: landscape and parliamentary enclosure in North-West England*. Lancaster: Centre for North-West Regional Studies, 2005.

Whyte, I. 'Parliamentary Enclosure and Changes in Landownership in an Upland Environment: Westmorland, c.1770–1860', *Agricultural History Review* 54 (2006): 240–56.

Williams, L.A. *Road Transport in Cumbria in the Nineteenth Century*. London: George Allen & Unwin, 1975.

Winchester, A.J.L. *Landscape and Society in Medieval Cumbria*. Edinburgh: John Donald, 1987.

Winchester, A.J.L. 'Wordsworth's "Pure Commonwealth"? Yeoman dynasties in the English Lake District, c.1450–1750', *Armitt Library Journal* 1 (1998): 86–113.

Winchester, A.J.L. *The Harvest of the Hills: rural life in northern England and the Scottish Borders 1400–1700*. Edinburgh: Edinburgh University Press, 2000.

Wood, J. 'Visualizations of Furness Abbey: from romantic ruin to computer model', *English Heritage Historical Review* 3 (2008): 8–35.

Woods, A. *A Manufactured Plague: the history of foot-and-mouth disease in Britain*. London: Earthscan, 2004.

Wordsworth, W. *Guide through the District of the Lakes ...* 5th edition. Kendal: Hudson & Nicholson and London: Longman & Co., 1835.

Wynne, B. 'Misunderstood Misunderstandings: social identities and public uptake of science', *Public Understanding of Science* 1 (1992): 281–304.

Yee, C. *The Silent Traveller: a Chinese artist in Lakeland*. London: Country Life, 1937.

Index